04-DEC-6

MODELING ONSITE WASTEWATER SYSTEMS AT THE WATERSHED SCALE:

A USER'S GUIDE

by:

John E. McCray
Hydrologic Science and Engineering Program, Colorado School of Mines

Mengistu Geza
Environmental Science and Engineering Division, Colorado School of Mines

Kyle E. Murray
Center for Water Research, University of Texas at San Antonio

Eileen P. Poeter
Geology and Geological Engineering Division, Colorado School of Mines

David Morgan
Oregon Water Science Center, U.S. Geological Survey

2009

WERF

The Water Environment Research Foundation, a not-for-profit organization, funds and manages water quality research for its subscribers through a diverse public-private partnership between municipal utilities, corporations, academia, industry, and the federal government. WERF subscribers include municipal and regional water and wastewater utilities, industrial corporations, environmental engineering firms, and others that share a commitment to cost-effective water quality solutions. WERF is dedicated to advancing science and technology addressing water quality issues as they impact water resources, the atmosphere, the lands, and quality of life.

For more information, contact:
Water Environment Research Foundation
635 Slaters Lane, Suite 300
Alexandria, VA 22314-1177
Tel: (703) 684-2470
Fax: (703) 299-0742
www.werf.org
werf@werf.org

This report was co-published by the following organizations. For non-subscriber sales information, contact:

IWA Publishing
Alliance House, 12 Caxton Street
London SW1H 0QS, United Kingdom
Tel: +44 (0) 20 7654 5500
Fax: +44 (0) 20 7654 5555
www.iwapublishing.com
publications@iwap.co.uk

Library of Congress Catalog Card Number: 2008940348
Printed in the United States of America
IWAP ISBN: 978-1-84339-528-7/1-84339-528-2

Colorado School of Mines

The research on which this report is based was developed, in part, by the United States Environmental Protection Agency (EPA) through Cooperative Agreement No. X-830851-01 with the Water Environment Research Foundation (WERF). However, the views expressed in this document are solely those of Colorado School of Mines and neither EPA nor WERF endorses any products or commercial services mentioned in this publication. This report is a publication of WERF, not EPA. Funds awarded under the Cooperative Agreement cited above were not used for editorial services, reproduction, printing, or distribution.

ACKNOWLEDGMENTS

The research team appreciates the assistance of Clifford "Kip" Bossong of the U.S. Geological Survey in Denver, Roy Laws of Jefferson County Department of Health and Environment (Colorado), Barbara Rich of the Deschutes County Environmental Health Division (Oregon), and Tiffany Yesavage at Colorado School of Mines, and the Curtis Cude at the Oregon Department of Environmental Quality. These persons and organizations provided data used in this research.

Research Team

Principal Investigator:
John E. McCray, Ph.D., P.G.
Hydrologic Science and Engineering Program,
Environmental Science and Engineering Division, Colorado School of Mines

Project Team:
Mengistu Geza, Ph.D.
Environmental Science and Engineering Division, Colorado School of Mines

David Morgan, M.S.
Oregon Water Science Center, U.S. Geological Survey

Kyle E. Murray, Ph.D.
University of Texas at San Antonio

Eileen P. Poeter
Geology and Geological Engineering Department, Colorado School of Mines

WERF Project Subcommittee:

Ken Jones, Ph.D.
Green Mountain Institute for Environmental Democracy

W. Keith Taulbee, Ph.D.
U.S. Environmental Protection Agency

Lilit Yeghiazarian, Ph.D.
Department of Bioengineering, University of California, Los Angeles

Leanne Whitehead
Tennessee Valley Authority

Water Environment Research Foundation Staff

Director of Research: Daniel M. Woltering, Ph.D.
Senior Program Director: Jeff Moeller, P.E.

ABSTRACT AND BENEFITS

Abstract:

This research provides guidance for conducting watershed-scale modeling assessments and decision making associated with onsite wastewater system (OWS) pollutants, with a focus on nitrogen and phosphorus. The researchers discuss fundamental modeling concepts and philosophy and summarize fundamentals of hydrology and pollutant transport related to OWS. They describe mass-balance screening models and GIS screening models, as well as complex numerical models that include groundwater models, vadose-zone models, surface-water models, and integrated watershed models. They provide guidance on model selection, obtaining model-input and calibration data, model parameterization, model-sensitivity analysis, model calibration, and long-term model care. They describe use of model results for risk-based decision making. Case studies demonstrate how the methodologies presented in the guide are applied, and demonstrate the use of models to evaluate alternative watershed-management scenarios. Reviews of GIS-screening models and watershed-scale models are included in the appendix.

Planners, regulators, modelers, OWS professionals, and hydrologists will find useful information in this guide. It is particularly useful to engineers who need to implement models for quantitative evaluations of OWS-related problems at the watershed scale, as well as to those people who seek to understand how models are used for watershed assessments and decision making.

Benefits:

- Provides guidance for watershed modeling associated with onsite wastewater.
- Explains the concept of watershed modeling and decision making for various users that include planners, regulators, engineers, and scientists.
- Enables rigorous and defendable quantitative assessments of onsite wastewater systems at the watershed scale.
- Describes how one can use model results for risk-based decision making.
- Presents rigorous model implementation procedures, including obtaining input data, model sensitivity analysis, and model calibration.
- Provides real world examples of using model results to evaluate different watershed-management scenarios and for decision making.

Keywords: Watershed modeling, decision making, nitrate, phosphorus, watershed management, septic systems, WARMF, Soil Water Assessment Tool.

TABLE OF CONTENTS

LIST OF TABLES

LIST OF FIGURES

CHAPTER 1.0

PURPOSE AND MOTIVATION

1.1 Purpose of the Guide

This user's guide provides guidance on modeling *watershed-scale* problems associated with decentralized wastewater-treatment systems (DWTS), with a particular focus on onsite wastewater systems (OWS). The guide focuses on modeling transport and fate of the nutrients nitrogen (N) and phosphorus (P) because these are the most common OWS constituents of concern, and because these pollutants are regulated in surface waters (N and P) and in groundwater (N). However, the information on presented in the guide on model construction, sensitivity analyses, calibration, risk analysis, and scenario evaluations is also relevant for modeling other pollutants, such as pharmaceuticals, pesticides, and microbial pollutants.

Decision makers can use the guide to determine whether relatively simple screening models (presented in Appendix A) are sufficient for use in the decision-making process, or if sophisticated models (presented in Appendix B) are more appropriate. The guide provides advice about the type of model that should be used for particular scenarios, and the data requirements for model implementation.

Modeling experts will particularly appreciate the guidance on important issues such as conceptual-model development, mathematical-model selection, model-sensitivity analyses, model uniqueness, and calibration.

Finally, the guide provides some real-world and hypothetical case studies that demonstrate the usefulness of using watershed-scale models, and provide templates for certain common scenarios relevant to the decentralized wastewater treatment community.

It was necessary to use specific models to develop this guide. The researchers tried to select the tools they thought were best-suited to OWS problems, and that were publicly available. However, it is not the intent of the guide to provide information relevant only to specific models, because this would be too restrictive. New models are developed regularly, and existing models are continually updated. Particular scenarios or existing expertise may influence the choice of a specific model. Thus, the guide provides advice relevant to classes of models, while relying on specific models to demonstrate important practical concepts associated with model use, and to simulate case studies of relevant watershed-scale scenarios relevant to OWS.

This guide focuses on the principles of watershed modeling, including model setup, calibration, and scenario evaluation that are applicable to most watershed models. Most of the frequently used physically based watershed models account for similar transformation and transport processes. Models may differ in the degree of parameterization; however, some of the input data and parameters are common to all. So, it is worthwhile to discuss the general procedures rather than focusing on a particular model. The discussions in this user's guide are not intended to lead model users to a particular model nor provide guidance to set up and run models for the reasons described below.

- Model Graphical User Interfaces (GUIs) are continuously updated. GUIs are altered several times and guidance on how to use a model may not be valid after some period. The researchers suggest that readers may consider a different type of model that has not been discussed in the guide.
- Model routines might be updated or new routines may be included. For example, Watershed Analysis Risk Management Framework (WARMF) (Chen et al., 2001) is currently more appropriate for OWS than Soil and Water Assessment Tool (SWAT) (Neitsch et al., 2005) because it has a module to directly address OWS inputs. However, an effort is currently underway to add an OWS module to SWAT, and SWAT may have more appropriate contaminant transport equations. Thus, model users will have options to use SWAT, WARMF or another watershed model for implementing OWS related studies. More information about SWAT and WARMF can be obtained using the information given in Appendix C.
- We believe that model users should opt for a model that is more robust and appropriate for the intended purpose. However, the modeler may be familiar with a particular model and may want to modify and implement this model for the sake of cost or time efficiency.

It would be more useful to present the various modeling systems that have been developed and used to answer a wide range of environmental questions including OWS as it is presented in the review of models than discussing a particular model. Modelers may find other similar watershed models, which might be appropriate. An inventory of available models that evaluates the models across a set of key characteristics is provided in Appendix B of this guide including their advantages and disadvantages.

1.2 Need for the Guide

Increasingly, planners, regulators, and other decision makers are considering the watershed-scale impacts of development and other land-use decisions, rather than only evaluating potential or existing impacts at the site and local scale. The U.S. EPA is promoting a watershed-scale approach with respect to permitting, mitigating, apportioning, or evaluating pollutant and nutrient loading from various sources. Thus, watershed-scale models of varying complexity are important tools for making quantitative assessments. Models, when applied properly, provide less subjective and more transparent information to aid decision makers. Models can be very useful in the evaluation of the relative impact of different land uses on surface-water and groundwater quality at local and watershed scales.

Watershed-scale models have been increasingly applied to agricultural problems in the last decade. Mathematical models are frequently used to evaluate groundwater contamination problems at scales of a few square kilometers or less, usually for groundwater plumes that evolve from hazardous waste sites or landfills. However, models have rarely been applied to investigate important scenarios associated with OWS. Models can be quite useful, however, when decision makers assess the relative risk to water quality associated with scenarios such as: allowing a large development to use individual or multi-housing OWS; planning for future land-use where OWS are involved with one or more land uses; or evaluating whether advanced treatment of N or P is warranted (i.e., additional treatment beyond that provided by a conventional OWS).

Mathematical models can also help determine potential contributions of OWS to pollutant loading in natural water systems relative to other sources. Currently, decision makers usually assume that OWS will degrade water quality. This could lead to overly restrictive requirements on homeowners or developers. Alternately, some hydrologic systems are more sensitive than others to potential impact from OWS, and thus decisions based on prior experience or existing water quality in an adjacent watershed may not be appropriate.

Soil has an inherent ability to effectively treat wastewater and reduce or remove potential pollutants, but this treatment capacity varies with soil type and hydrologic conditions. Clearly, not all watersheds have equal capacities to attenuate or treat potential pollutants associated with OWS. It is difficult to provide robust and reliable information to decision makers regarding the potential impact of OWS on water quality without using quantitative tools.

Complex models can be expensive to implement, partially because of the labor cost for experts that can implement models, but also because of the cost of acquiring the appropriate data to build reliable models. As a general rule, simpler models discussed in Chapter 5.0 are less costly but yield results that are much more uncertain. They are also less defendable because they generally do not represent actual conditions with respect to hydrology or pollutant fate and transport. Thus, screening models and tools that implement simplifying and conservative assumptions should be used to first evaluate if OWS impact is a potential concern within the limits of uncertainty associated with the model results. If OWS impact is shown to be a reasonable concern, then models with increasing complexity and implementation cost can be used in a sequential manner until the decision makers are comfortable with the model predictions, including the model uncertainty, relative to the risk associated with the decisions.

This guide demonstrates the potential benefits of using mathematical models to assess potential impacts of OWS on water quality at the watershed scale, and provides guidance on how to choose and apply models of varying complexity.

1.3 Audience for the Guide

1.3.1 Primary Users

This guide is intended primarily for:

- planners/regulators who need to decide whether to implement a quantitative (modeling) solution to a watershed problem related to OWS, and model type to use;
- professionals who will implement quantitative GIS methods to evaluate OWS impacts at the watershed-scale; and
- professional hydrologic modelers hired to implement watershed models.

1.3.2 Secondary Users

The guide will also be useful for:

- OWS engineers and scientists who are not modelers;
- hydrologists, environmental scientists, and GIS specialists who are not modelers, but who want to learn more about OWS and quantitative methods that assess OWS-related issues at the watershed scale.

The next chapter, "How to Use this Guide," directs each type of user to the appropriate sections, and suggests a reading order of the relevant sections to optimize the guide's usefulness.

CHAPTER 2.0

HOW TO USE THIS GUIDE

The information in this guide is useful to planners and regulators (Section A), technical persons who are not hydrologists (Section B), hydrologists who are not modelers (Section C), and modeling experts (Section D) who may or may not be familiar with OWS applications.

This chapter directs each type of professional to the relevant chapters, using the links provided in Table 2-1 below. Some sections include information that is redundant to other sections; this is necessary so each type of user can most effectively use the guide.

All users should read Chapter 1, which describes the purpose of the guide. All users should also consider Chapter 3.0, the "primer" for OWS. Even those who are familiar with OWS from a regulatory, planning, or traditional testing perspective will likely find useful the information related to fate and transport of onsite contaminants in the environment. Users who are well versed on OWS as well as on the physics, chemistry, and biology associated with fate and transport of nitrogen, phosphorus, pharmaceuticals, other organic contaminants, and microbial contaminants, can skip Chapter 3.0. The first part of Chapter 10.0 on risk based modeling approaches should be interesting to all readers, while some sections of this chapter are more relevant to modelers who are implementing the approach. Chapter 11.0 describes the case studies, and would be of interest to all readers.

Table 2-1. Relevant Reading Material for Various Users.

Reader	Most Relevant Chapters								
Planners & Regulators	C3	C4	C5	C6				C10	C11
OWS Technical Persons who are not Modelers	C3	C4	C5	C6	C7			C10	C11
Modelers not familiar with Watershed Modeling	C3			C6	C7	C8	C9	C10	C11
Watershed Modelers who are not familiar with OWS applications	C3					C8	C9	C10	C11

2.1 Chapters Relevant to Planners and Regulators

Planners and regulators should start with Chapter 3.0 for an overview of fate and transport of OWS contaminants. This information provides the basis for all quantitative approaches, whether these are simple screening models or complex numerical models (discussed in Chapter 5.0).

Chapter 4.0 includes information on how and why we model. This chapter includes a discussion of what models actually are, what they can and cannot do, some common modeling myths, model terminology that is useful for understanding the remainder of the guide (and for becoming more conversant with those eccentric modelers), our general philosophical approach to modeling, an important discussion on how models can be used to assess the risk associated with a particular decision, and a summary of a case study which briefly illustrates some concepts put forth in this chapter.

Planners and regulators who feel very knowledgeable about models should at least read the latter sections of Chapter 4.0, starting with section H (the modeling approach of parsimony, or starting simple). In particular, we find that most of us sometimes forget the importance of assessing uncertainty in models and data used to construct or evaluate models. Chapter 4.0 serves as a reminder that all decisions, whether made with the help of a model or not, has some associated risk or uncertainty that we cannot always anticipate. Models, of course, can be highly useful in quantifying this risk in a more objective, consistent way.

Chapter 5.0 will also be useful to decision makers. This section explains the different types of models, the varying complexity of model classes, how they represent hydrological systems, and when their use is appropriate.

Chapter 6.0 is a guide to model selection, including a decision matrix that will step the Wthrough the process of deciding whether a model is appropriate, and choosing the appropriate model for a particular decision.

Chapters 7.0 through 10.0 (model input parameters for various model types, model calibration and evaluation, risk-based decision making using uncertain models) are directed mainly toward hydrologists and modelers who are implementing a model for a watershed-scale OWS problem. Most of the basics associated with these chapters that are most useful to planners and regulators are described in Chapter 4.0 (how and why we model). However, sections A through C in Chapter 10.0 discuss the general aspects and basis behind using models for a risk-based approach, and will probably be useful to most planners and regulators.

Chapter 11.0 describes case studies of modeling approaches used for planning-and-management scenario evaluations and decision making. This chapter is a detailed summary of the most important features of the case studies. The detailed case studies in their entirety, or information on how to retrieve the complete information on a case study, are included in Appendix C.

2.2 Chapters Relevant to OWS Technical Persons who Are Not Hydrologists or Modelers

Technical professionals tasked with evaluating watershed-scale impacts of OWS, but who are not hydrologists or hydrologic modelers by training, may want to review Chapter 3.0 for an overview on fate and transport of OWS contaminants. This information provides the basis for all quantitative approaches described in this guide, whether these are simple screening models or complex numerical models (see Chapter 5.0).

Chapter 4.0 includes information on how and why we model. This chapter includes a discussion of what models actually are, what they can and cannot do, some common modeling myths, model terminology that is useful for understanding the remainder of the guide (and for becoming more conversant with those eccentric modelers), our general philosophical approach to modeling, an important discussion on how models can be used to assess the risk associated with a particular decision, and a summary of a case study which briefly illustrates some concepts put forth in this chapter.

Chapter 5.0 will also be useful to technical professionals. This section explains the different types of models, the varying complexity of model classes, how they represent hydrological systems, and when their use is appropriate.

Chapter 6.0 is a guide to model selection, including a decision matrix that will step the user through the process of deciding whether a model is appropriate, and choosing the appropriate model for a particular decision.

Chapter 7.0 discusses data needs for various models, and how to obtain data that are readily available and that are usually not calibrated (climate, shallow soil properties, elevation, etc.). Technical personnel should pay particular attention to the information in this chapter when evaluating the level of effort that is likely required for a quantitative evaluation (time, budget, additional field monitoring, etc). It will be particularly useful to identify data gaps in the early stages of the watershed evaluation.

Chapters 8.0 through 10.0 (model input parameters for various model types, model calibration and evaluation, risk-based decision making using uncertain models) are directed mainly toward watershed modelers who are implementing a model for a watershed-scale OWS problem. However, sections A through C in Chapter 10.0 discuss the general aspects and basis behind using models for a risk-based approach, and will probably be useful to most OWS professionals.

Chapter 11.0 describes case studies of how modeling approaches can be used for planning-and-management scenario evaluations and decision making. This chapter is a detailed summary of the most important features of the case studies. The detailed case studies in their entirety, or information on how to retrieve the complete information on a case study, are included in Appendix C.

2.3 Chapters Relevant to Modelers Who Are Not Watershed Modelers

Modelers who are tasked with evaluating watershed-scale impacts of OWS, but who are not familiar with OWS, should review Chapter 3.0 for an overview of OWS. In addition, you may find useful the information about the fate and transport of certain OWS contaminants for which you are not familiar (basic concepts related to nitrogen, phosphorus, organic contaminants including some pharmaceuticals, and microbial contaminants are discussed).

Chapter 5.0 explains the different types of models, the varying complexity of model classes, how they represent hydrological systems, and when their use is appropriate. This may be useful to you if you have mainly focused on one type of model (e.g., MODFLOW) for a very long time.

Chapter 6.0 is a guide to model selection, including a decision matrix that will step the user through the process of deciding whether a model is appropriate, and choosing the appropriate model for a particular decision, including simple screening models.

Chapter 7.0 discusses data needs for various models, and how to obtain data that are readily available and that is usually not calibrated (climate, shallow soil properties, elevation, etc.). Technical personnel should pay particular attention to the information in this chapter when evaluating the level of effort that is likely required for a quantitative evaluation (time, budget, additional field monitoring, etc). It will be particularly useful to identify data gaps in the early stages of the watershed evaluation.

Chapters 8.0 through 10.0 (parameterizing a model, model calibration and evaluation, risk-based decision making using uncertain models) are directed mainly toward modelers who are implementing a model for a watershed-scale OWS problem.

Chapter 10.0 is new territory for most modelers in the environmental consulting arena. This chapter discusses various methodologies for model uncertainty analyses that are all based on probability distributions of uncertain input parameters. The methods vary in complexity and effort, and include calculation of prediction intervals based on inferential statistics, targeted simulations that consider specific cases that are agreed upon by the stakeholders, to stochastic Monte Carlo simulations.

Chapter 11.0 describes case studies that demonstrate how modeling can be used for planning-and-management scenario evaluations and decision making. This chapter is a detailed summary of the most important features of the case studies. The detailed case studies in their entirety, or information on how to retrieve the complete information on a case study, are included in the Appendix C.

Review of watershed models is also included in the user's guide. Appendix A is a "Review of Mass-Balance and Other GIS-Based Screening Models for Watershed-Scale Assessments of Onsite Wastewater System Impacts,"and Appendix B is a "Review of Potential Distributed Models for Watershed-Scale Assessments of Onsite Wastewater System Impacts."

2.4 Chapters Relevant to Watershed Modelers Not Familiar with OWS

Watershed modelers who are not familiar with OWS should review Chapter 3.0 for an overview of OWS. In addition, some of the information about the fate and transport of certain OWS contaminants can be useful if the modeler has no prior experience with these contaminants (i.e., nitrogen, phosphorus, organic contaminants including some pharmaceuticals, and microbial contaminants).

The experienced modeler will already be familiar with the concepts in Chapter 4.0. However, it is useful to become familiar with the terminology used in this guide. In addition, modeling professionals who are not familiar with uncertainty analysis as applied to risk assessments and decision making will find this section useful. The last section briefly describes a case study that illustrates some concepts put forth in this chapter, which will help the modeler appreciate the view and philosophy held by the authors of this guide.

Chapter 6.0 is a guide to model selection, including a decision matrix that will step the user through the process of deciding whether a model is appropriate, and choosing the appropriate model for a particular decision that is being considered. Most experienced modelers are well versed with the concepts in this chapter.

Chapter 7.0 discusses data needs for various models. This includes the conventional sources for physical and chemical hydrologic data, as well as some creative sources for data specifically related to OWS applications at the watershed scale. Most material on physical and chemical hydrologic data are familiar territory to the seasoned modeler. However, the information on obtaining data relative to OWS sources, as well as for some types of models that the modeler has not previously used, may be particularly useful and insightful.

Chapters 8.0 through 10.0 (parameterizing a model, model calibration and evaluation, risk-based decision making using uncertain models) are directed mainly toward modelers who are implementing a model for a watershed-scale OWS problem.

Chapter 8.0 describes the hierarchical process of how model input parameters are obtained for watershed-scale models. Those with experience in modeling, but new to watershed-scale modeling with multiple hydrologic compartments, or who have experience with physical hydrologic modeling rather than chemical transport modeling, may be surprised at the large number of input parameters required for watershed-scale OWS modeling. Thus, it is inevitable that some parameters will be obtained via calibration, some will be measured, some will be estimated independent of calibration, and some will be fixed using literature values, or allowed to vary within probabilistic distributions of values obtained from the literature. Much of this parameter determination process depends on the sensitivity of the model to input parameters, whether input-parameters are correlated, what data are available or can be obtained at reasonable cost, and whether prediction or uncertainty analysis is desired.

Chapter 9.0 describes state-of-the-art methodologies for model calibration that are currently transitioning from the research arena to the real-world. This chapter is likely to be useful to most hydrologic modelers. Chapter 9.0 has a particular focus on multiple-parameter, spatially distributed, transient watershed-scale models, but the principles are relevant to any hydrologic model, for one or integrated hydrologic compartments, and for simple or highly complex model types discussed in Chapter 5.0. The chapter discusses issues related to model sensitivity, parameter correlation, uniqueness of the model solution, and how all these issues relate to calibration and model performance evaluation. Methods for evaluating prediction uncertainty are discussed. The importance of and suggested methodology for long-term model-performance evaluations, which include post-audits and monitoring programs, are presented in the last two sections.

Chapter 10.0 is new territory for most modelers in the environmental consulting arena. This chapter discusses various methodologies for model uncertainty analyses that are all based on probability distributions of uncertain input parameters. The methods vary in complexity and effort, and include calculation of prediction intervals based on inferential statistics, targeted simulations that consider specific cases that are agreed upon by the stakeholders, to stochastic Monte Carlo simulations.

Chapter 11.0 describes case studies that demonstrate how modeling approaches can be used for planning-and-management scenario evaluations and decision making related to watershed-scale OWS issues. This chapter is a detailed summary of the most important features of many different case studies. The detailed case studies in their entirety, or information on how to retrieve the complete information on a case study, are included in the appendices.

Review of watershed models is also included in the user's guide. Appendix A is a "Review of Mass-Balance and Other GIS-Based Screening Models for Watershed-Scale

Assessments of Onsite Wastewater System Impacts" and Appendix B is a "Review of Potential Distributed Models for Watershed-Scale Assessments of Onsite Wastewater System Impacts."

2.5 Important Documents External to the User's Guide

Appendix C is a list of "Important Documents Associated with this Guide" and information on how to retrieve them. The guide often refers to important documents that cannot be reproduced here because of copyright laws, or because updates to the document are frequent and cannot be accounted for in this guide. These documents are pointed out in the text, and compiled in Appendix C. Appendix C also contains information on how a user can obtain the documents.

CHAPTER 3.0

PRIMER FOR ONSITE WASTEWATER SYSTEMS (OWS) AND TRANSPORT OF CONTAMINANTS IN THE ENVIRONMENT

3.1 Basics of OWS

3.1.1 Septic Tanks, Septic Tank Effluent

Decentralized wastewater systems (i.e., septic systems) serve approximately 25% of households in the U.S. (U.S. EPA, 2003) and contribute about 3.78 million ML/year of wastewater to the shallow subsurface (Frost et al., 2002).

3.1.2 Infiltration of STE into the Subsurface

Septic systems are designed to leach wastewater (i.e., effluent) into the shallow subsurface or vadose zone (Murray et al., 2007). This design allows for biodegradation in the soil and unsaturated zones to reduce potential contaminants to acceptable levels before the septic tank effluent (STE) recharges an underlying aquifer.

3.1.3 The Receiving Environment

Onsite wastewater systems (OWS) integrate treated effluent into the environment by directly discharging the effluent into surface water or by infiltrating the effluent through a soil treatment unit (STU). For most nutrients or compounds, partial or complete removal from the aqueous phase is expected as water infiltrates through the vadose zone. Mechanisms for removal are different for each compound and include irreversible sorption to soil, aerobic biodegradation, anaerobic biodegradation, volatilization, and abiotic degradation (ex.: hydrolysis). Some compounds may not be treated effectively because of the geochemical composition or microbial population present in the soil/vadose zone; hence, these compounds may be detected in groundwater and surface waters (Drewes et al., 2003; Hinkle et al., 2005; Zimmerman, 2004).

3.2 Pollutant Transport Processes

3.2.1 Transport and Transformation of Nitrogen

Nitrogen as an element can be found in nature in a wide range of valence states, from the oxidized N^{+5} to the reduced N^{-3}. This enables nitrogen to form a variety of inorganic and organic compounds (presented in Table 3-1). Because of the presence of N in many compounds, it can be utilized by microbes as an electron donor or acceptor in several oxidation-reduction reactions (Figure 3-1). Oxidation-reduction reactions and other processes involving N include the following:

- Ammonification (or "mineralization"): breaking down of organic nitrogen compounds, such as amino acids, to ammonium with no valence change.

- Nitrification: oxidation of inorganic nitrogenous compounds such as ammonium or ammonia. Nitrification is considered a two-stage process, each stage is governed by a different group of microbes. *Nitrosomonas* convert ammonium to nitrite, a short-lived intermediate nitrogen species, which in turn is converted to nitrate by *nitrobacter*. Both processes require oxygen.
- Denitrification: reduction of oxidized nitrogenous compounds such as nitrate or nitrite to a gaseous phase. The denitrification chain is expressed as:
  $NO_3^- \rightarrow NO_2^- \rightarrow NO \rightarrow N_2O \rightarrow N_2$, although N_2 is not necessarily the end-product in all denitrification processes. The oxygen atom released in each stage is combined with carbon that is released from the breakdown of organics to form CO_2.
- Nitrogen fixation: some microbes are capable of turning the usually-inert gaseous N_2 into ammonium or simple amino acids.

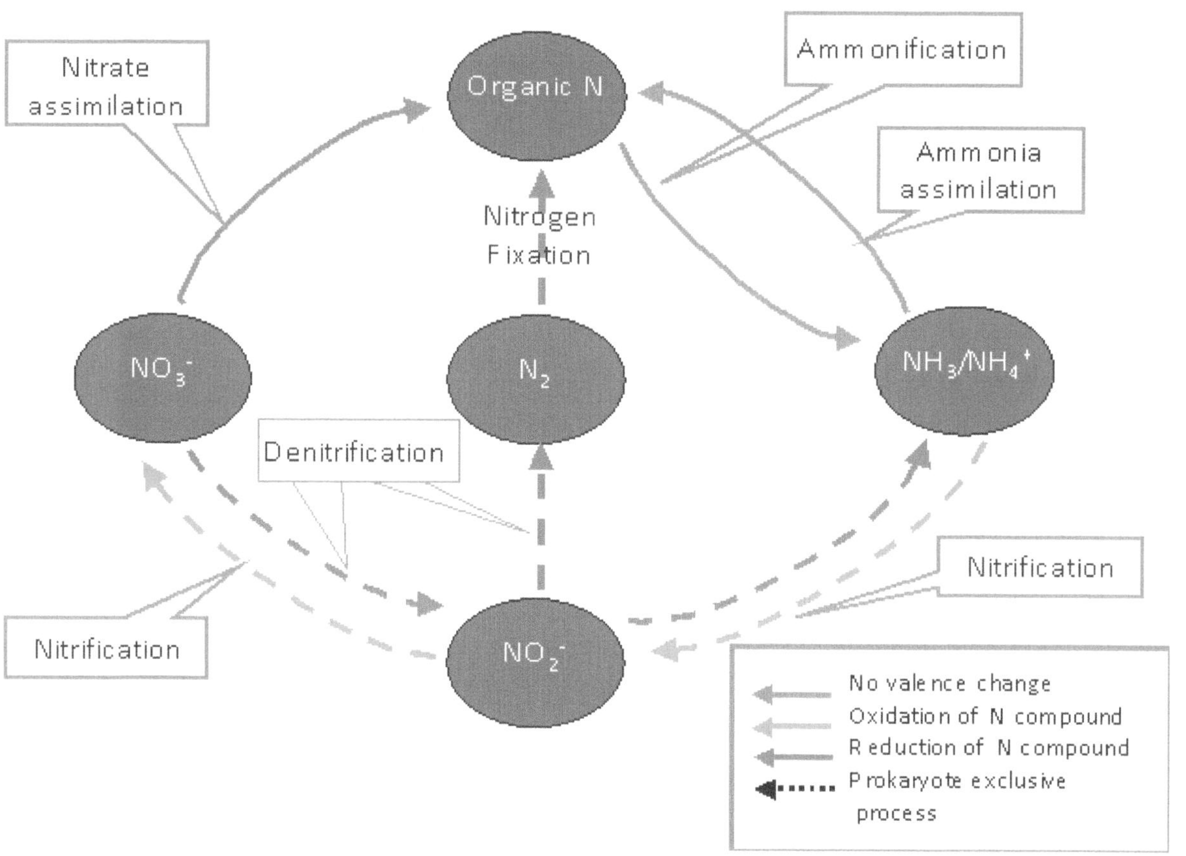

Figure 3-1. The Nitrogen Cycle in Nature, Main Nitrogen Species and Processes.

Table 3-1. Common Nitrogen Species and Their Chemical Compositions.

Compound	Chemical Composition	Remarks
Dinitrogen gas	N_2	
Nitrous oxides	NO, N_2O, N_xO	Green-house gases
Ammonia / ammonium	NH_3 / NH_4^+	Relative concentrations of these two are pH dependant
Nitrite	NO_2^-	
Nitrate	NO_3^-	
Amino acid	$NH_2CHRCOOH$	R refers to an organic group and is different for every amino acid

While nitrogen acts as a valuable nutrient, it can also become a pollutant when released in large quantities to the environment. For example, nitrate in drinking water has been associated with methemoglobinemia, which affects the ability of red blood cells to bind to oxygen (Shuval and Gruener, 1972). Infants are at greater risk when drinking NO_3^--rich water, hence the common name given to methemoglobinemia: the "Blue Baby Syndrome." The United States Environmental Protection Agency (U.S. EPA) and the World Health Organization (WHO) have set a limit of 10 mg NO_3^--N/L in drinking water (Bedessem et al., 2005). Nitrate is also an important and limited nutrient in oceanic environments (Ryther and Dunstan, 1971). Discharge of NO_3^--rich groundwater to surface waters (fresh or salt water) can lead to eutrophication – algal blooming – and deterioration of water quality (Brandes et al., 1974; Weiskel and Howes, 1992).

On the watershed scale, all these concerns are related. A massive rain event in the upper Mississippi river, for example, can increase the leaching of nitrates from OWS or fertilized agricultural fields to groundwater reservoirs used as drinking waters. Flooding of fertilized fields may release nitrates to runoff water and to the Mississippi river. Where the Mississippi waters are discharged to the Gulf of Mexico, hundreds or thousands of miles away from the rain event, a "dead zone" is created in the gulf waters due to eutrophication. The algal bloom causes depletion of oxygen ("hypoxia"), which leads to mortality of fish and other marine organisms (Howarth et al., 2006). The damage is both ecological and financial, as fish and shellfish production is impacted.

3.2.1.1 Nitrogen in the Atmosphere

Nitrogen in the gaseous form of N_2 is the most abundant gas in the atmosphere, and comprises about 78% of all gases. In this form, nitrogen is inert and is not utilized by most living organisms, apart from specific groups of microbes that are able to fixate the inert gas and produce organic nitrogen compounds. The chemical properties of N_2 make its utilization less common than other diatomic gases because microbes must convert N_2 into an organic molecule by overcoming the strong covalent triple-bond between the two nitrogen atoms. The enzyme responsible for catalyzing nitrogen fixation has high activation energy, and requires an environment depleted of oxygen (Lodwig et al., 2003).

Nitrous oxides are highly reactive gases that may be produced in the process of denitrification and contribute to environmental pollution. First, NO and N_2O can act as components of acid rain, in the form of nitric acid. NO is a green-house gas, about 300 times more active than CO_2. As N_2O moves up in the atmosphere into the stratosphere, it may participate in reactions that result in the destruction of ozone (Brady and Weil, 2002).

Nitrogen losses to the atmosphere in the form of ammonia can be appreciable in certain applications. Algae that live in fish ponds or wetlands such as rice paddies extract CO_2 from the water in the process of photosynthesis. That reduces the amount of carbonic acid and drives up the pH of the aqueous environment, commonly above pH 9. Under such basic conditions the ammonium ions in solution are naturally converted to ammonia, which volatilizes to the open air. Septic tank effluents may be a bit on the basic scale, but are a minor source of ammonia volatilization, mostly due to their underground discharge.

3.2.1.2 Nitrogen in Vegetation

Nitrogen is important to all life since it is an essential component of amino acids, which are the building blocks of proteins. Nitrogen is crucial for the assemblage of chlorophyll in plants. A nitrogen-rich soil stimulates root growth and plant productivity, and the uptake of other nutrients. Overall, healthy plant foliage can contain between 2.5% and 4% nitrogen (Brady and Weil, 2002). Excess nitrogen, however, may cause late maturity in plants, over-growth that leads to relatively weak stems, and sensitivity to fungal and insect attacks.

There are several ways that a plant obtains nitrogen:

- One of them is through mutual symbiosis with the *rhizobia* bacteria, which create nodules in legumes roots (Figure 3-2) and are able to fix atmospheric nitrogen. The plant provides the bacteria in an oxygen-depleted environment and energy in the form of dicarboxylic acid, produced in the process of photosynthesis, and in return the bacteria releases ammonia or amino acids. The biological reduction of atmospheric N_2 to ammonium provides about 65% of the biosphere's available nitrogen (Lodwig et al., 2003).
- Plants that do not live in such cooperation have to rely on nitrogen supplied from the soil pore water. The plant roots take up nitrate and ammonium ions. (Usually, though, nitrite is not of concern because of its trace quantities in the soil). These ions become available to the plant through natural decomposition of organic matter in the soil, or through nitrogen fertilization in agricultural applications.
- Plants are also capable of sorbing ammonia from the atmosphere. Forests may act as a major sink for ammonia carried by wind from fertilized crop lands many miles away.

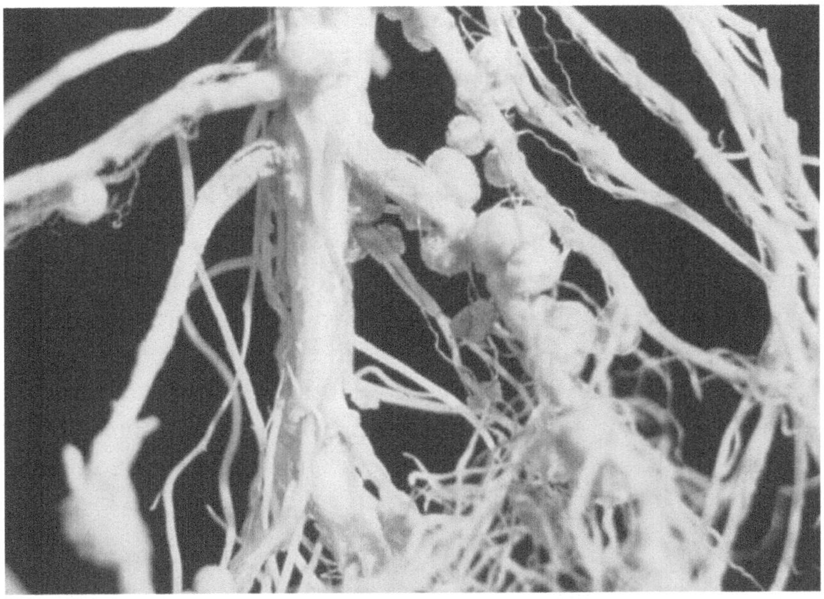

Figure 3-2. Rhizobium Nodules on Roots of Legumes (soils.usda.gov/sqi/concepts/soil_biology/bacteria.html). Printed with permission from SAFS Group at University of California Davis.

3.2.1.3 Nitrogen in Soils and Groundwater

Possible sources of nitrogen in soils include the following:

- Atmospheric deposition of nitrogen: nitrous oxides released to the atmosphere through denitrification and fossil fuel combustion are turned into nitric acid when dissolved in water, as during rainfall (Brady and Weil, 2002).
- Nitrogen fixation: conversion of diatomic nitrogen gas to organic matter by specialized soil bacteria (Lodwig et al., 2003).
- Wastewater application: animal waste is commonly used as a natural and cheap fertilizer. Human waste is released to soils through OWS in about 25% of the homes in the U.S. (U.S. EPA, 2002). The total nitrogen concentration in STEs varies from system to system, but is usually around 68 mg/l (McCray et al., 2005). The dominant nitrogen species in STE are ammonium (~85%) and organic nitrogen (~15%), with trace amounts of nitrate and nitrite. Wastewater plants may also use soil infiltration as their final wastewater purification process (Diab and Shilo, 1988).
- Fertilization: artificial nitrogen fertilizers have been applied to agricultural soils ever since the early 20th century when a process was developed to enable nitrogen to be cheaply synthesized into ammonia. Fertilizer made from this ammonia is estimated to be responsible for sustaining roughly 40% of the world's population and is the source for 40-60% of the nitrogen in the human body (Fryzuk, 2004).
- Geological sources: dinitrogen gas comprises 78% of the atmospheric gases, but the atmospheric reservoir is only 2% of the nitrogen that exists on our planet. The rest is buried in rocks, and is released to the environment through erosion and volcanic activities (Brady and Weil, 2002).

3.2.1.4 Nitrogen Transformation

Denitrification The denitrification process, as described in "main nitrogen processes," is carried out by heterotrophic bacteria. Heterotroph is a general name for forms of life that need an organic source of carbon to survive (humans included), unlike autotrophs, which are able to create organic carbon compounds from inorganic carbon (CO_2). There are several requirements for the denitrification process to occur.

- Oxygen depletion: some denitrifiers are facultative (obligatory) anaerobic in genera, and require an anoxic environment. Such environments can be achieved when common heterotrophic bacteria, or nitrifying bacteria, consume the dissolved oxygen from the pore water, and oxygen diffusion back into the water is limited. Oxygen dissolution into the pore water can be influenced by soil texture, and limited when the soil pores are small and diffusion of gaseous oxygen into the pores is slow. Saturation of the soil can also reduce the rate of oxygen diffusion in the soil pore water, since there is less water-air interface to allow transfer of oxygen from the air into the water. Low oxygen levels are considered beneficial to denitrification when the soil pores are at least 60% saturated, or when the soil air contains no more than 10% oxygen (often related) (Brady and Weil, 2002). However, denitrification is believed to occur also in well-aerated soils, in anaerobic microsites (McCray et al., 2005).
- Carbon source: when conditions are anoxic, nitrate is used as a final electron acceptor in the respiration process, instead of oxygen. In the denitrification process, organic matter is oxidized and carbon dioxide is released, much like in an oxygenic respiration process. Substantial denitrification may occur in "carbon hot-spots" in the soil, where decomposing roots or fauna act as carbon sources (Parry et al., 2000). In OWS the organic carbon source is primarily the wastewater.
- pH: the optimum denitrification pH is 7-8 (Martin and Focht, 1977).
- Temperature: the optimum temperature for denitrification is 25-35°C, but the process will occur between 2 and 50°C (Brady and Weil, 2002).

Ammonium Adsorption To understand the conditions that allow or interfere with nitrogen leaching from soils to groundwater, one must first understand the affinity of the soil components to attract or repel nitrogen ionic species.

Under neutral pH, most soil particles exhibit a negative net charge on their surfaces (Sposito, 1989). That allows the attraction of soil particles to positively charged ions, such as hydrogen ions (H^+) and ammonium (NH_4^+). This immobilization of ions on soil particle surfaces is called adsorption, and is dependent on the charge and size of the ions in soil pore water, the charge of the soil particles and the available adsorption sites on the soil grains' surfaces. The amount of adsorption sites depends on the grain size (generally, the smaller the grains in the soil, the more adsorption sites exist) and the grain mineralogy. When the pH of the soil is reduced, hydrogen ions, which are positively charged, are attracted to the negatively charge surfaces of soil particles, and reduce their ability to adsorb more positively charged ions. On the other hand, at low pH, due to saturation of hydrogen ions, the soil may have a greater affinity to attract negatively charged ions.

Ammonium, being a positively charge ion, has a good potential to adsorb to soil particles, and the potential for of ammonium to leach into groundwater is low. Because of its small size, ammonium can become entrapped within cavities in the crystal structure of certain clays (Brady and Weil, 2002). Soils with high amounts of clay minerals of the type 2:1 have a high capability of preventing ammonium mobilization in soil solution, as well as soils rich in humus (organic matter). In highly weathered (tropical) soils ammonium sorption is minor because little 2:1 clay is present. While adsorbed, the ammonium ions are held in exchangeable form, available for plant uptake (though slowly released), and partially protected from leaching.

Nitrification Apart from sorption, ammonium in the soil can undergo nitrification. Nitrification is the biological oxidation of ammonia with oxygen into nitrite followed by the oxidation of these nitrites into nitrates. Nitrification is an important step in the nitrogen cycle in soil. Nitrification plays an important role in the removal of nitrogen from wastewater because the conventional removal process is nitrification, followed by denitrification.

Nitrate is highly mobile in soils under neutral-pH conditions, due to its negative charge. The nitrate ions are repelled by most particle surfaces, which makes the nitrate leaching potential high. When nitrate reaches groundwater, concentrations are reduced mainly due to dilution. In some hydrological systems, mostly in temperate regions with plenty of rainfall, groundwater is discharged to streams or lakes (the groundwater is "feeding" the surface body of water) (Figure 3-3). The anoxic conditions that prevail in the riparian zones (the contact between land and water) and in river or lake sediments have a high denitrification potential (e.g., Mengis et al., 1999), along with plant uptake.

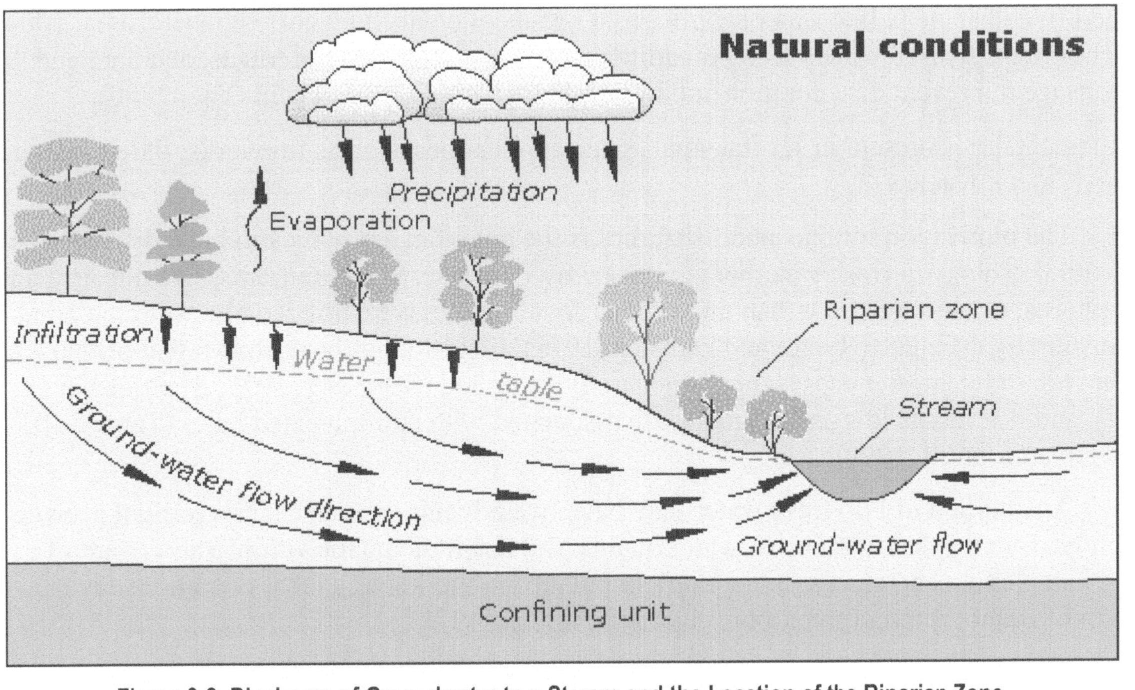

Figure 3-3. Discharge of Groundwater to a Stream and the Location of the Riparian Zone.
(source: http://ga.water.usgs.gov/edu/earthgwdecline.html)

3.2.1.5 Transport and Transformation of Phosphorus

Phosphorus is commonly found in the following forms:

- Orthophosphate: an inorganic series, which includes H_3PO_4, $H_2PO_4^-$, HPO_4^{2-}, and PO_4^{3-} species. The transition between these species is pH dependent, with $H_2PO_4^-$ and HPO_4^{2-} the dominant orthophosphate species in neutral-pH waters.
- Organic P: includes P incorporated with organic compounds, such as sugars, phospholipids, and nucleotides. Adenosine-triphosphate, or ATP, is a phosphorous-bearing organic molecule that serves as an energy storage unit in living cells. Some synthetic organic compounds, including some insecticides, contain reduced forms of P.
- Phosphorus minerals: these are reactions of phosphate with different cations (mostly iron, aluminum, calcium, and lead) forming minerals such as hydroxyapatite, fluorapatite, strengite, vivianite and more. The source of these minerals is usually magmatic rocks, yet they can be found in sedimentary environments (e.g., marine environments with high productivity, coastal environments enriched with bird droppings [guano]) where there is enough P and suitable cations. Hydroxyapatite is also the major constituent of enamel, which is the hard outer part of the tooth. Minerals that contain calcium and P (calcium phosphates) are commonly used as fertilizers.
- Condensed phosphates are derived mostly from detergents and cleansers, for water "softening" (removing of divalent cations such as calcium and magnesium to prevent precipitation of carbonates). They include polyphosphate forms such as $P_2O_7^{4-}$ and $P_3O_{10}^{5-}$, and degrade, or hydrolyze, slowly to orthophosphate (see above). Polyphosphates are an additive in some public water supplies as a means of controlling corrosion (Rezania, 2004). They are also used widely as food additives in cheese, fish paste products, and ham and sausage to prevent discoloration and to stabilize vitamin C (Sekiguchi et al., 2000).
- Elemental P is present in several small-scale applications such as fireworks, flares, napalm and safety matches.

The motivation for phosphorus studies is the potential eutrophication and disruption of the natural ecology in fresh waterbodies caused by P release. Some nutrients, like nitrogen and phosphorus, occur at no more than micromolar levels and may be utilized almost to the point of exhaustion by the algae (Ryther and Dunstan, 1971). Field studies have shown that primary production in freshwater ecosystems is generally P-limited (Schindler, 1977; Miettinen et al., 1997), and P levels as low as 0.03 mg/L are associated with eutrophicated lakes (Table 3-2) (Newton and Jarrell, 1999).

An addition of P to fresh water may have two adverse effects: excessive algal growth, which leads to the deterioration of water quality and death of aquatic wildlife as a result of oxygen depletion; and change in the nutrient balance in the water, so that conditions favor growth of cyanobacteria (green-blue algae) (Schindler, 1977).

When the N-to-P ratio (N:P) in a fresh waterbody becomes low (~5), due to an addition of P or depletion of N, algal competition for N is usually dominated by species that are capable of fixating N from the inert N_2 gas) (Gerritse, 1993). The cyanobacteria group has the N fixation capability, and also tend to release toxins to the water as secondary metabolism (metabolism by-product). Many cases have been reported on poisoning of vertebrates and invertebrates following ingestion of cyanobacterial bloom / scum material, and at least 60 different cyanobacterial toxins have been recognized (Codd, 1995).

Table 3-2. Typical Total P Concentrations (mg/L) by Trophic State (Oligotrophic = lowest productivity, Hypertrophic = highest productivity).
Adapted from Newton and Jarrell (1999)

Aquatic System	Trophic State			
	Oligotrophic	**Mesotrophic**	**Eutrophic**	**Hypertrophic**
Lake	<0.01	0.01 – 0.03	0.03 – 0.1	>0.1
River	---	<0.01	0.01 – 0.05	>0.05

Phosphorus in Vegetation Phosphorus is usually a limiting nutrient for plants in most soils. The tendency of phosphate to bond to aluminum and iron oxides and hydroxides, and the precipitation of phosphate minerals such as hydroxyapatite, make it relatively unavailable for plants (Brady and Weil, 2002).

Plants receive their phosphorus through root uptake. The plant roots uptake mainly inorganic dissolved phosphorus (phosphate ions), yet are capable of uptaking some soluble organic phosphorus compounds. Phosphate ions are also delivered to plant roots through a mutualistic symbiosis with mycorrhizal fungi (Brady and Weil, 2002; Powell and Daniel, 1978). These fungi are microscopic, thread-like, and act as root extensions, as they reach into the soil several centimeters from the plant roots. The fungi are able to absorb phosphorus ions from soil solution, and may even be able to access some strongly bound forms of phosphorus. The phosphates are transported into the root via the fungi cells, thus bypassing possible interaction of phosphate with the soil particles. The plant then incorporates most of the phosphorus into plant tissue, and some is used for synthesis of adenosine triphosphate (ATP) molecules, deoxyribonucleic acid (DNA) and ribonucleic acid (RNA).

Generally, the phosphorus content of a healthy leaf tissue is 0.2-0.4%. When phosphorus supply is limited, the plant is able to transfer phosphorus from older leaves to the newer, rapidly growing leaves (Brady and Weil, 2002).

Phosphorus in Soils The concentration of phosphorus in a soil solution is very low, usually ranging from 0.001 mg/l in very infertile soils to about 1 mg/l in heavily fertilized soils. The main sources of phosphorus in soils are:

- Decomposition: organic matter is broken down by microbes, which utilize part of the released organic phosphorus for biomass building. Phosphorus is recycled in the soil following cell death and decomposition.
- Fertilization: because phosphorus is a limiting nutrient, farmers add chemicals that contain phosphates, or animal manure, which naturally contains phosphorus. Cow manure can contain up to 6 g of phosphorus per kg (dry weight) of manure, while swine slurry may contain more than 4 times that concentration (Kleinman et al., 2002), yet the concentrations depend on the animals diet (Chapuis-Lardy et al., 2004). Most of the P is in the inorganic form (Chapuis-Lardy et al., 2004).

- Wastewater: as explained for nitrogen, OWS practices include the release of wastewater to the soil, as well as some wastewater treatment plants. Phosphorus forms common in domestic wastewater are orthophosphate, polyphosphate and organic P. Polyphosphate-containing detergents were previously a major source of phosphorus in domestic wastewater. Several states started to ban the use of phosphorus-containing laundry detergents in the mid-1970s, and as more states joined the ban the industry voluntarily stopped including phosphorus in household laundry detergents. However, phosphorus in dish-washing detergents is still common, and its use is not restricted (Litke, 1999).
- Dust: a small amount of phosphorus (0.05-0.5 kg per hectare per year) enters the soil from the atmosphere, sorbed on dust particles. This small amount nearly balances the losses from the soil in undisturbed forest and grassland ecosystems (Brady and Weil, 2002).

Phosphorus Transformation Phosphorus losses from soil pore water are a bit more complicated than nitrogen because some of the phosphorus reactions in the soil are reversible. Phosphorus is not generally lost from the soil in gaseous form (although some phosphorus may be lost as phosphine gas (PH_3) – a phenomenon noted in certain graveyard soils) (Brady and Weil, 2002). The a-biotic retentions of phosphorus in the soil can be described as adsorption and precipitation.

- Sorption: Phosphates can be adsorbed to certain particle surfaces, such as aluminum and iron oxides and hydroxides, clays with a 2:1 structure, and calcium carbonates. In acidic solutions, where the amount of H^+ ions increases, the ability of the negatively-charged phosphates to adsorb to mineral surfaces increases. Sorption and precipitation are often related. Phosphate can adsorb to about 5% of pure calcite (calcium carbonate) surfaces, despite the mineral's negative charge under slightly-basic to basic pH conditions. While only a small portion of the mineral's surface is available, these adsorption sites act as nuclei for the precipitation of calcium phosphate minerals, mostly at high P concentrations (Borrero et al., 1988). At low P concentrations, iron is still a more efficient adsorbent than calcium carbonate, whereas fine-grained calcium carbonate adsorbs P slightly more than clay minerals such as illite and smectite due to a low specific surface area (Borrero et al., 1998). Organic acids can compete with phosphates for sites on mineral and soil surfaces, thereby decreasing phosphate adsorption. Certain organic acids can entrap reactive aluminum and iron in stable organic complexes called chelates, thus making them unavailable for reaction with P. On the other hand, phosphates can be immobilized by organic carbon, as organic-bound complexes, and thus removed from solution. Overall, organic soils are not efficient in removing phosphates (Brady and Weil, 2002). When a soil solution changes from an oxidizing environment to a reducing environment, due to oxygen depletion, phosphates tend to be released back to solution, and their leaching potential increases (Zurawski et al., 2004).
- Precipitation: In calcareous soils and marine coastal environments, the most common P mineral precipitate is apatite ([$Ca_{10}(PO_4)_6X_2$], X=anions; mainly hydroxyl [OH^-] and fluoride [F^-]). Typical calcium sources for the precipitation reactions are: shell fragments in coastal marine environments (Whelan, 1988); in semi-arid and arid terrains a calcareous horizon is typically developed in the soil (Sposito, 1989); calcium carbonate can precipitate as a normal weathering product of calcium-bearing primary silicates, such as pyroxene, amphiboles and feldspars (Sposito, 1989); and calcium can also be found in soils developed from carbonate rocks (Borrero et al., 1988). Acidic pH encourages dissolution of calcium-carbonate and release of calcium ions into solution, which should theoretically contribute to the formation

of calcium phosphate minerals. However, hydroxyapatite [$Ca_{10}(PO_4)_6(OH)_2$] is more stable under basic conditions (Maurer et al., 1999).

Phosphate tends to bind to calcium cations in basic soils with high Ca/P ratio, but to iron, aluminum, magnesium and manganese cations in acidic soils. In non-calcareous terrain, the developments of acidic conditions cause gibbsite [$Al(OH)_3$] dissolution and, subsequently, variscite [$AlPO_4 \cdot 2H_2O$] precipitation (Zurawski et al., 2004). Calcium phosphate minerals such as hydroxyapatite [$Ca_{10}(PO_4)_6(OH)_2$] and fluorapatite [$Ca_{10}(PO_4)_6F_2$] have lower solubility products (i.e., less likely to dissolve and release phosphates back to the solution) than iron phosphate minerals such as strengite [$FePO_4 \cdot 2H_2O$] and vivianite [$Fe_3(PO_4)_2 \cdot 8H_2O$]. The formation of struvite [$MgNH_4PO_4$], for example, in wastewater is improbable due to its large solubility product and the fact that its equilibrium is rapidly achieved (Maurer et al., 1999). A solution can be supersaturated with regard to a mineral, due to a slow rate of mineral formation (Robertson et al., 1998).

It is believed that precipitation of phosphate minerals such as hydroxyapatite (HAP) is a two-step process. First, a surface complex is formed, which is not a pure compound under "dirty waste-water conditions" but an amorphous mixed crystal with a higher solubility product than HAP (Maurer et al., 1999). This surface complex has a relatively large solubility product. Second, over time the surface complex exothermically crystallizes into the less soluble HAP. There is a link between aerobic/anaerobic environments and the stability of phosphate minerals. In anaerobic (reducing) environments, the dominant iron form is the soluble Fe^{2+} (ferrous ion), whereas in oxidized environments the dominant form is the insoluble Fe^{3+} (ferric ion). Vivianite, which is composed of ferrous ions and phosphates, is a major P sink under reducing environments, with a lower solubility product under basic conditions (Robertson et al., 1998). In oxidizing environments the dominant iron phosphates are variscite or strengite, which have lower solubility products under acidic conditions (Brady and Weil, 2002).

CHAPTER 4.0

PRIMER ON WHY AND HOW WE MODEL

4.1 What are Models?

Readers who are not familiar with technical applications of models will find this chapter particularly useful. It provides an overview of how mathematical models of varying complexity can support assessments of the watershed-scale impacts of OWS pollutants. It explains what models can and cannot do, and discusses the use of models in the decision-making process. Additional details on each of these aspects are given in subsequent chapters. In conjunction with Chapter 2.0, "How to Use this Guide," readers will gain a better understanding of how the details fit together with the "big picture" of watershed-scale modeling.

Models incorporate processes of interest into a single framework that performs calculations using known information (inputs) to obtain desired information (outputs). Developing a "*conceptual* model" is the first step in the modeling process. This includes both the big-picture and detailed concepts.

A simple example is a model to balance one's checking account. The big-picture *conceptual* model involves the concept that we start with a certain amount of money in the bank, and we want to keep track of debits from the account so that we know how much money has been spent, and how much money remains. Some details of the conceptual model could include whether check or ATM cards are used, whether interest is applied to our balance, how the interest is compounded, etc.

For a watershed model, the conceptual model could include big-picture items such as whether the problem is related to aquifer groundwater quality or stream water quality. It would also include concepts such as whether an aquifer is comprised of sedimentary material or fractured rock; whether pollutant transport is primarily applied to a land-surface, to the subsurface, or both; whether snowmelt is an important process; whether a particular hydrologic process can be simplified or must be conceptualized as being very complex; whether phosphorus sorption or precipitation as an insoluble mineral dominates phosphorus removal in the soil, etc. An essentially unlimited number of examples are possible.

Mathematical models are quantitative implementations of conceptual models that provide numerical results (output) based on numerical data (input). However, mathematical models are only as good as the conceptual model upon which they are based. Thus, *conceptual models should be as detailed and rigorous as possible!* One can then better decide whether they should simplify – or even ignore – certain processes in the mathematical model.. At least, one can better assess the likely uncertainties that are brought about by the simplification or omission.

A simple example of a mathematical model is a spreadsheet that is set up to balance a check book. The inputs are the dollar values for each check or ATM withdrawal, and the initial condition regarding how much money is in the bank at the start of model implementation. The desired information, or output, is the updated balance of how much money is (or was) in the bank at any given time. The framework includes the calculations that subtract the values from the prior balance and computes a new balance, as well as the way the model is set up (e.g., the frequency of data input, and whether a balance is calculated after every check, or only at the end of the month).

A watershed model is similar in concept to a model that balances a checkbook. Indeed, most models are based on mass and energy balances. In fact, a watershed model can be almost as simple as the check-book model described above. However, it can also be very complicated and include thousands of equations and hundreds of different types of input data. The complexity of the model must match up with the type of information that is desired from the modeling effort. Of course, complex models are more difficult and expensive to implement.

Models are useful tools for watershed evaluation, planning and assessment. Models provide a means to incorporate what we know about a watershed into a mathematical framework to help in design and/or decision making. Models have long been used for various engineering and water-resources applications. For examples, they are used to design aircraft, bridges and cars; to optimize highway construction and traffic; to optimize placement of groundwater wells for delivering potable water; to contain and remediate contaminant plumes at hazardous-waste sites; for flood prediction and mitigation; to assess movement of agricultural pollutants in watersheds; and countless other applications.

In particular, with respect to watershed assessments and planning, models are typically used to *quantify* the impact of certain land-use or planning changes (e.g., using OWS in a watershed, re-zoning from agricultural to residential use, etc.) on watershed parameters of interest (e.g., stream flow, or water quality in streams or groundwater). These quantitative outputs can then be used by stakeholders and policy makers to aid in decision making, and/or to justify certain decisions. However, generally, models do not make decisions; only humans should make decisions and policy.

The goal of this user's guide is to provide guidance on model selection, implementation, and decision making for watershed-scale assessments related to OWS.

4.2 What Models Can Do

A model is a representation of a system based on understanding the types and magnitudes of relationships. It is an aid for evaluation and decision making that relies on the quality of inputs. If models are constructed and used responsibly, the sky is the limit with respect to applications. First and foremost, models provide information to aid in decision making. They serve as tools to estimate or predict the effects of changes in environmental conditions or land management within a watershed. Although models have limitations and uncertainties, they are still the best tools for making predictions despite uncertainties.

The following list is intended to be informative, not exhaustive, and to provide specific examples of how watershed-scale models can be used. Model user(s) expect that watershed models can:

Aid in understanding how a system "works"

Complex and simple models are useful tools to help users understand cause-and-effect relationships. For many hydrologic systems, the responses to perturbations in the input are not linear. That is, a five-fold change in input may only effect a 10% change in output. Conversely, a 10% change in some input could affect a five-fold change in model output for a particular parameter or process. For example, a model could be used to understand how lower nitrogen concentrations in septic tank effluent (e.g., relative to a potential rule or regulation change) would impact groundwater or stream concentrations.

Aid in evaluating the potential impact of different land uses

A model could be used to evaluate the relative changes in stream flow, constituent concentrations changes, or chemical mass loading for several different land-uses. Planners may wish to assess whether residential areas with OWS would contribute more or less nitrogen loading to an aquifer or stream than a current agricultural use. Planners may wish to evaluate the relative phosphorus loading that would be contributed to a stream from construction of a development versus the OWS that are associated with the development.

Aid in evaluating the potential impact of OWS on water quality

A properly constructed model can help assess eventual stream or groundwater concentrations of OWS wastewater pollutants, as well as the timing of "release" should any impacts occur.

Help evaluate the potential watershed-specific impacts of OWS

Regulations related to OWS and water quality are often issued county-wide or state-wide. However, potential impacts are likely to differ considerably between different watersheds, regardless of geopolitical boundaries. Thus, a solution that may not work in one watershed may be appropriate in another watershed. Models can facilitate decisions that are more tailored to the characteristics or needs of a particular watershed.

Help evaluate the benefits and disadvantages between using OWS or sewers and wastewater treatment plants (WWTPs) for wastewater treatment

A model could be used to assess whether sewers and wastewater treatment plants would significantly reduce the concentrations of constituents-of-concern in groundwater or streams. In many cases, OWS can result in water quality that is favorable compared to WWTP discharges.

Help evaluate the impact of OWS on water quantity

The user can evaluate how much infiltrated wastewater is lost through evapotranspiration, and the impact on the water levels in underlying aquifers. If drinking water is delivered via a central distribution system, but wastewater is treated via OWS, then the potential increase in local groundwater levels or stream flow can be evaluated.

Aid in development design

Development design factors such as lot size or even housing unit size can be evaluated. The model can be used to maximize development while minimizing soil, groundwater, or stream concentrations of a particular constituent.

Aid in cost analysis

Using an example from above, the model could be used to see how the various concentrations that might be discharged in septic tank effluent from OWS outfitted with engineered treatment systems (often called advanced treatment units) actually influence groundwater or stream concentrations. In this manner, the benefit of the additional costs of these systems can be evaluated.

Screen worst-case scenarios

Simpler models (discussed in Chapter 5.0) can be used to determine whether impacts from OWS are likely under worst-case scenarios (e.g., no mixing or attenuation). This type of model can help determine whether more resources should be devoted to data gathering or toward development of more complex models or quantitative tools.

Help develop TMDLs

The U.S. EPA regulates watershed and stream loading based on total daily maximum loads (TMDLs), where stakeholders limit loadings from many different sources. A model can help planners evaluate how different watershed loading plans might influence TMDLs.

Assist with evaluating specific areas that are most likely impacted by OWS pollution

A model can be used to assess which parts of an aquifer, or what portions of a stream are most likely to be impacted by OWS pollutants, and which areas are likely to not experience impacts.

Help design and optimize watershed monitoring and data collection

A thoughtfully constructed watershed model can make clear what additional data are most useful to generate a rigorous quantitative assessment of OWS impacts at the watershed scale (for present or future conditions). Thus, limited resources can best be focused on the most important data types and locations.

Help users understand the uncertainty associated with a quantitative watershed assessment

All watershed assessments are associated with uncertainty. Fortunately, a model can be used to quantify the uncertainty. Thus, a model can be very useful to help the user to understand the uncertainty and use this information to make an informed, risk-based decision.

Serve as a powerful tool to evaluate the risk associated with a planning or land-use decision

Models can be used to assess the risk associated with the implementation of a certain land-use decision, even if the model output is uncertain. A likely range of outcomes can be calculated using a model, which can be used to develop a probability or frequency associated with different outcomes. Then, planners and regulators can use this information to make decisions based on their willingness to accept the quantified risks. The use of models to assess risk is discussed in more detail in Chapter 10.0.

4.3 What Models Cannot Do

One part of understanding modeling is appreciating its limitations. A model is only as good as the modeler's knowledge of the system used to construct the model and the data supplied to run the model. A model is not a replacement for understanding a system, is not independent of experts, and is not a substitute for good science and field work.

A system with very little overall understanding of watershed functions and limited data is not likely to be accurately modeled. Models sometimes are perceived as "black boxes" because the assumptions, uncertainties, and methods are not clearly identified. Without clearly identifying what factors contribute to the development of the model, there won't be much public trust and confidence in the results. With that said, there is not enough justification to simply resist models, especially when considering their benefits. However, model users should not expect watershed models to:

Make land-use decisions or implement policy

Technical people hired to implement models can provide information to decision makers based on model calculations; but only stakeholders, regulators, and policy makers can make decisions.

Exactly represent nature

Models are mathematical representations of a watershed, and even the most complex models are simplifications of a real watershed. Cost and time constraints associated with watershed characterization and data collection necessitate simplification. In addition, some pollutant-transport or hydrologic processes are likely occurring that are unknown to engineers and scientists who develop the models.

Provide exact predictions

Models can be very useful for making predictions. However, one cannot expect a model to exactly reproduce stream or groundwater concentrations, or flow in a stream, at some point in the future. If the model was properly calibrated and evaluated, then model results are likely to result in predictions that closely match measured conditions.

Be appropriate for all OWS-related watershed applications

Most models are designed for a specific purpose (e.g., for runoff associated with agricultural use, for pollutant transport to and in streams, contaminant transport in aquifers, etc). One model that is commonly used by a particular engineer or consulting firm may not be appropriate to use for a particular OWS-related problem. Thus, the watershed scenario to be evaluated and the model used must be matched up carefully.

Be used for one watershed if it is developed for another watershed

Hydrology and pollutant transport is likely to differ considerably between watersheds. Thus, a specific model that is constructed for one watershed should not be used in another watershed, even if the watersheds are thought to be similar.

Remove uncertainty from land-use decisions

Even the most carefully constructed, calibrated, and evaluated model will have uncertainty associated with the results. However, planners and regulators must be prepared to accept that using a model will not remove the uncertainty from a problem.

Be used to plan or predict periods for significantly longer than the calibration period

Often the stresses in the watershed hydrologic system that occur after calibration are significantly different than those that occurred during the calibration period. A general rule is that a model can be used to make predictions for a time period that is equivalent to the time period from which data was used to calibrate the model. If the model is to be used for decisions that are applicable over a longer period, then a model uncertainty analysis should be conducted that considers the likely long-term changes in the modeled system.

Continue to be applicable over the long run if post-modeling evaluations are not made

To be most effective, models that are used for long-term should be continually evaluated and updated beyond the first calibration and application of the model. Stresses to the natural system (e.g., different hydrologic stresses and patterns, as well as land-use changes) that were not considered during model development may occur after the model is constructed. Additional data may also be available that was not available during model construction and initial evaluation. Thus, it is not usually reasonable to expect model predictions to be valid if model performance is not continually evaluated. Unfortunately, these post-audits, as they are called, are not frequently conducted.

4.4 Common Modeling Myths

Myth #1: Models are Too Expensive to Implement

Models can be time consuming and costly to implement. However, the cost of model development to aid a decision-making process should be weighed against the potential benefits of the model, the total cost associated with the decision, and the risks of making an uninformed decision.

For example, the Heatwole and McCray (2007) listed in Appendix C implements a complex vadose-zone fate and transport model to provide information on likely nitrogen concentrations in shallow groundwater that would result if a 1000-home development used OWS to treat domestic wastewater. This model was used effectively to evaluate the risk of nitrogen contamination, and to aid in the final decision on how to proceed with the development. The cost of model implementation and associated data collection totaled about $30,000. The total worth of the development (i.e., the selling price of all homes in the development) would be about *half a billion* dollars. The tax base generated by these homes would be millions of dollars *per year*, and the profit of the developer would be *hundreds of millions* of dollars. The costs of installing advanced nitrogen removing treatment units for every home in the development to mitigate potential nitrogen release into the environment would likely cost more than *one million* dollars. The costs of removing nitrogen from groundwater in the future would also likely exceed *several million* dollars.

Under these circumstances, it is difficult to argue that the $30,000 price tag on modeling was too expensive. Rather, the problem is that the costs of such assessments are not included in the planning and budgetary process.

The authors hope that this user's guide can help educate planners, developers, and regulators on the usefulness of modeling assessments so that the costs of such evaluations will be considered in the planning stages of a project, rather than in the regulatory or damage-control stages.

Myth #2: If a Model Says it's True, it Must be True (Or, Model Output is Always Correct)

The output of a model is only as correct as the conceptual model upon which the mathematics are based, and on the accuracy of the input data. In truth, no model can be a close representation of reality. However, the representation can be sufficient for the model to be useful.

Myth #3: The More Complex the Model, the Better It is

We adhere to the principle of parsimony when it comes to modeling. This principle essentially means that the model should be as simple as possible while still including the relevant hydrologic or pollutant-transport features that are deemed important. This principle is attributed to 14th century English philosopher William of Occam, who contended that the best explanation is the simplest explanation. This approach has been referred to as "Occam's razor" because the explanations are shaved down to the essentials. Einstein amended this philosophy for models by saying that "models should be as simple as possible, but no simpler." Indeed, adding unwarranted complexity makes the model more difficult to parameterize and calibrate, and makes it more likely that a model will not have a unique result.

The model should be designed to provide output that produces information relevant to the problem at hand, and minimize output that is not relevant. Then, the model formulation and input should be simplified as much as possible to enable calculation of the desired output, while still retaining the relevant features of the watershed system to be modeled. If appropriate data cannot be obtained to parameterize and calibrate a more complex model, then a simpler model that can be parameterized and calibrated may be necessary.

Finally, it is always best to start with simpler models that might be useful for screening. If the screening model is chosen to be conservative (typical), then the modeling might stop after its application because the model suggests that the undesired effect will not occur. If the screening model result suggests an undesired effect, then a slightly more complex model might be implemented that has less simplified assumptions. Ultimately, a complex numerical model might be required to incorporate all the aspects of the system that are desired.

Myth #4: Using a Model Eliminates Human Error

Human errors can be made in model selection, in choosing model boundary conditions, in the setup of the model (i.e. translating the physical characteristics into the mathematical model), in measurement or reporting of data used for model input or calibration, in the process of model calibration, in interpretation of model results, and in other areas. In short, human error can give modeling a bad name.

Myth #5: Models are Objective, and Can Remove the Subjective Nature of Decision Making

Models can never be truly objective because they involve decisions made by humans related to the underlying conceptual picture upon which the model is based. Humans decide which equations to use, which data to utilize, and how to mathematically simulate the natural system. Most of these decisions are subjective to varying degrees.

Models can lead to a less subjective decision-making process, however, because the often complex concepts are implemented in an efficient, repeatable, systematic and documented framework. In addition, models are generally based on accepted fundamental principles related to water and pollutant movement in our environment. Thus, complex decisions made with the aid of models are usually more defendable than decisions that are made in a less systematic or consistent fashion.

Myth #6: Literature Data can be Readily Used for Model Input

Most parameters related to hydrology and pollutant transport in watersheds are highly variable. Thus, any particular literature value that might be obtained, especially if it was measured or obtained for a different watershed, is most likely not appropriate for use. Ideally, watershed specific data should be used for all input parameters; however, this is not typically realistic. Modelers should determine a reasonable range of uncertainty for the unknown parameters (the literature is very useful for this task), and can evaluate the resulting uncertainty in the model.

Myth #7: Default Parameters in a Model are Reliable for Most Model Applications

Default values for model input parameters are often obtained for an early application of the model, or are determined based on the judgment of the model developers. Using default values without evaluating whether they are applicable for the current scenario is rarely appropriate. In some cases, default values truly are widely applicable. (Examples could be the physical constants, or parameters that do not vary significantly on a global scale). As a general rule, users should carefully evaluate all model-input parameters and try to obtain values that are appropriate for the watershed or sub-watershed being modeled, or at least endeavor to understand and quantify the uncertainty associated with using a default parameter.

Myth #8: Graphical User Interfaces are the Best Way to Execute a Modeling Project

Graphical user interfaces (GUI) help the modeler to more efficiently implement a model, but may limit the modeler's understanding of how the model works. In addition, using the GUI may severely restrict troubleshooting. The modeler should have the ability to manipulate the model outside of the GUI during the latter phases of modeling.

4.5 Model Terminology Used in this Guide

The number of terms that could potentially be defined in this section could extend into the hundreds. The researchers have limited the terms to those that will best help the practitioner "speak the modeling language."

Hydrologic simulation models – models that "simulate" the real world of hydrology and chemical transport using mathematics. These models use mathematical equations to calculate results for variables of interest such as stream flow, runoff volume, groundwater flow, pollutant concentrations in streams, lakes and groundwater, etc. These models can be classified as either theoretical (physically-based) or empirical models.

Run time, or simulation time – the real time required to conduct a model simulation. Spreadsheet calculations can require less than tenths of seconds. Some complex watershed or vadose zone models can require hours to days of real time to complete the simulation.

Watershed model – simulates the processes of interest for an entire watershed, which usually includes detailed representations of water and chemical transport in climate, runoff, infiltration, groundwater storage, and stream flow. These models usually do not have a rigorous representation of vadose-zone or the aquifer processes. Examples include WARMF, SWAT, HEC, etc.

Groundwater model – simulates the processes of interest for an aquifer or a selected portion of an aquifer, which usually includes detailed representations of water and chemical transport in the aquifer. These models usually do not have a rigorous representation of vadose-zone, surface-water, or climate processes. The most-used groundwater model is the MODFLOW package.

Stream model – simulates the processes of interest for a stream or a stream segment, which usually includes detailed representations of water and sediment transport, and may include chemical transport and some runoff processes. These models usually do not have a rigorous representation of vadose-zone, surface-water, or climate processes. Examples include HEC-HMS, TR55, SWMM or QUAL-2E.

Empirical model –not based on theoretical scientific principles, but rather on an equation or system of equations that can reproduce a desired observed output based on known inputs without regard to the physical principles. Regression models are common examples. Some empirical relationships have become so established that they are now regarded as "laws." A good example is the Darcy equation that relates groundwater flow to changes in hydraulic head.

Theoretical (physically based) model – uses general laws or theoretical principles. If all the governing physical laws were well known and could be described by equations of mathematical physics, chemistry, and biology, the model would be physically based. However, all existing theoretical models simplify the physical system and often include obviously empirical components (e.g., the conservation of momentum equation used to describe surface flow includes an empirical hydraulic resistance term).

Event model – represents a single event occurring over a relatively short period of time (e.g., ranging from about an hour to several days). A runoff event is a typical example of such a model. The initial conditions in the watershed for each event must be assumed or determined by other means and supplied as input data. The accuracy of the model output may depend on the reliability of these initial conditions.

Continuous (transient) model – operates over an extended period of time to calculate a time-variant parameter (flow, concentration, etc). At the beginning of the run, all initial conditions must be known or assumed. However the effect of the selection of those initial conditions decreases rapidly as the simulation advances.

Steady-state model – calculates the long-time, steady flows or concentrations. In truth, most hydrologic models never achieve "steady state," so this model represents an approximation in most cases. Steady-state simulations can be achieved by running transient models until the output is steady. Often, models are designed to calculate steady-state conditions (that is, the transient parts of the output are not considered). This can be advantageous (particularly when conducting model calibration) because such a model is only required to perform all calculations once with respect to time, and thus it runs much more quickly than a transient model.

Calculation unit – often also called the model element. This is the smallest fundamental *spatial* element within which all relevant calculations are performed. For example, a certain process may require 12 different equations to describe hydrology, pollutant transformation, and movement of water and chemical to the adjacent calculation units, and to calculate hydraulic head, chemical concentrations, etc. All 12 equations would be solved in each calculation unit, and a value for the desired variable (head or concentration) is calculated in each unit. A model may use one or millions of calculation units to represent a watershed or aquifer.

Time step – transient models can provide output for all spatial model elements at specified times during the simulation. Equations are solved and calculations are performed in every calculation unit, for every time step. Watershed and aquifer models commonly have dozens to thousands of time steps.

Lumped-parameter models – do not explicitly take into account the spatial variability of inputs parameters or outputs. They are usually structured to utilize average values of the watershed characteristics. Averaging a certain parameter also implicitly averages the process being represented. Typically, a model is called a lumped parameter model if one calculation unit is used for the entire watershed, aquifer, or other area of interest. Of course, ignoring the spatial variability can lead to significant errors, but is often a good approach for a screening model.

Distributed models – include spatial variation in inputs and output parameters and variables. Each calculation unit represents a model element. Groundwater models are often divided into cells that are similar in size and shape, but may be made smaller in order to incorporate desired changes in subsurface properties (e.g., heterogeneities). Distributed watershed models typically use topographic catchments as the calculation unit. Models have often been mistakenly classified as "lumped," even though they can represent spatial variability by subdividing the basin into segments with representative "lumped" parameters for each segment (e.g., HSPF, WARMF). A word of caution on terminology: no model can be discretized such that each model cell has truly unique hydrologic or chemical properties. Even distributed models used lumped parameters at some scale.

Analytical model – exactly solves a hydrologic or pollutant transport equation. To solve the equations almost always requires the user to accept a highly simplified conceptual model. For example, it is not usually possible to account for hydrologic variables or properties that change with time or space. Examples that cannot be considered include variations in precipitation or OWS input rate, groundwater flow rate or direction, spatial variations in hydraulic or chemical reaction properties, changing hydraulic stresses, or complex hydrogeologic or chemical boundary conditions. Sometimes, analytical models can account for simple heterogeneities, such as layered aquifers or soil.

Numerical model – is capable of solving the more complex equations that describe groundwater flow and solute transport. These equations generally describe multi-dimensional groundwater flow, solute transport and chemical reactions, although there are one-dimensional numerical models. Numerical models use approximations (e.g., finite differences, or finite elements) to solve the differential equations describing groundwater flow or solute transport. The approximations require that the model domain and time be discretized. In this discretization process, the model domain is represented by a network of grid cells or elements, and the time of the simulation is represented by time steps.

Deterministic model –uniquely specifies (or determines) all input parameters and produces a single output. Stochastic (probabilistic) model - attempts to account for the unknown variations in input parameters, and incorporate this uncertainty in the calculated output. Model input parameters are not described by unique values, but rather by probability distributions. Similarly, outputs are generally presented as probabilistic distributions. Examples of simple to complex probabilistic modeling approaches are provided below.

Screening model – is any model used to assess whether additional analysis is necessary. A screening model can identify potentially sensitive areas or highlight areas where more data are needed, or quickly test alternative conceptual models. Consequently, screening models provide cost-effective evaluations on strategies before more expensive physically-based models are employed for a specific site.

Continually stirred mixed reactor (CSTR, or mixing) model – is typically used for pollutant transport. The model accounts for time-variant changes in concentration input to the system, within the system, and leaving the system. The CSTR assumes a hydrologic compartment (usually a lake or aquifer), or reactor, is continually "well mixed." The output concentrations typically decline or increase exponentially depending on whether input concentrations are greater than or less than the resident concentrations in the hydrologic compartment. Some watershed models use this approach to transfer chemicals between model elements.

Spreadsheet model – is any model that can be implemented using a spreadsheet. In most cases, the spreadsheet model is relatively simple, although some can be quite complex. These models are relatively easy to use and readily accepted, in part because most professionals are familiar with spreadsheets.

Geographic information system (GIS) – is a computer system for capturing, storing, manipulating, and displaying geographically referenced data.

GIS screening model – combines a simple conceptual model with analytical equations that can be solved using a GIS. It generally captures the regional spatial characteristics of a system without including the detailed processes or data density that would be necessary for site-specific delineations. When integrated within GIS, computations become powerful and time-efficient because of advantages in data storage, manipulation, analysis, and visualization of spatial data.

Calibration – varying one or more input parameters until model output matches measured data that are set as calibration targets. Calibration can take dozens or even hundreds of model runs, and can be quite time consuming. For this reason, consider using one of several automated software packages that are available.

Calibration targets – data that are used for model calibration. Usually, the data are collected over a time period of several months to years for a watershed model, and are often collected at different locations within a watershed. Examples include hydraulic head measured in aquifers, stream flow, chemical concentrations measured in groundwater, surface water, or soil water above the water table.

Inverse modeling – the process of obtaining input parameters based on matching model output to measured data, rather than independently determining model-input parameters and running the model forward. This topic is described in more detail in Chapter 9.0 on calibration.

Sensitivity analysis – the process of varying model-input parameters in a systematic manner and quantifying the resulting change in model output. The purpose of this analysis is to understand which input parameters have the most important impact on model output. This can help the modeler to assess which input parameters should be measured or rigorously estimated, which model-input parameters can be fixed at reasonable values based on experience or on a literature search, and which parameters need to be carefully evaluated via an uncertainty analysis if rigorous measured data are not available. A good sensitivity analysis can take hundreds of model runs, and can be quite time-consuming. For this reason, consider using one of several automated software packages that are available. These topics are described in more detail in Chapters 9.0 and 10.0.

Prediction uncertainty analysis – neither the model output nor experimental data (input data or calibration targets) are known with certainty. Methods exist to express model calculations as an expected value with associated confidence limits. This topic is described in more detail in Chapter 9.0 and 10.0.

Monte Carlo uncertainty analysis – the process of varying the unknown or uncertain input parameters within a realistic probable range to understand and quantify the resulting uncertainty associated with the model prediction. This type of analysis is often conducted in lieu of model calibration (i.e., when it is not time- or cost-effective to collect the appropriate data). This method is primarily used for two purposes: 1) to statistically quantify model-prediction uncertainty when model-input parameter values describing spatial characteristics of the basin are not known or are highly uncertain; and 2) to address model-prediction uncertainty associated with the uncertain future scenarios of stress on the system (e.g., to make predictions related to changing climate). The model provides a range of probable outputs that, in turn, represent a range of probable predictions. In Monte Carlo analyses, a range of potential (yet reasonable) values could be obtained for most model input parameters. Then, an "ensemble" of model simulations are run where the value for each input parameter for each simulation is obtained via random sampling (often structured to minimize the number of runs required while capturing the distribution of the input parameters) from the uncertain input-parameter distributions. The resulting ensemble of model outputs represents a range of predicted outcomes. The model user then decides if s/he wants to rely on the median model prediction, or some other cut-off within the model-output distribution, to base a decision. The variability in model predictions can be very large when using this approach. The Yucca Mountain project relies heavily on Monte Carlo analyses due to a scarcity of measured data for most hydrologic parameters, and the need to provide predictions tens of thousands of years into the future. However, a simpler and less costly methodology for uncertainty analysis may be applied for OWS scenarios. The researchers termed this type of analysis a "targeted probabilistic uncertainty analysis," which is defined below. Uncertainty analysis is described in more detail in Chapters 9.0 and 10.0.

Targeted probabilistic uncertainty analysis – this approach is similar to Monte Carlo analyses in that the main purpose is to understand the potential variability in model prediction when the model input is uncertain or unknown. This approach is generally taken when model calibration is not possible or is not rigorous due to sparse data. For OWS applications, it is anticipated that this would mainly be performed in the pollutant-transport phase of the modeling because data for calibration are less likely to be available as compared to hydrologic data. In this approach, a model sensitivity analysis would first be conducted to determine which input parameters are most sensitive. Then, probable ranges for each of these important input parameters would be developed, ideally as a cumulative frequency diagrams (CFD), where the median as well as other percentile values for the uncertain parameters, could be determined. Then, model simulations are run using specific percentile values for each input parameters that are agreed upon by various stakeholders. The uncertainty in model prediction is evaluated by the stakeholders to help them arrive at a decision. This is similar to Monte Carlo analysis except sampling of input variables is limited and chosen by stakeholders.

Model uniqueness – some input parameters are correlated such that several different combinations of values for the sets of input parameters may result in the same model output result. However, only one of these combinations could be theoretically valid. Models that are not unique are not likely to provide good results beyond the time period for which the model was calibrated or evaluated.

Model error – mistake in the calculated model output that is due to an error in the conceptual or mathematical model. Examples include errors that result from assuming incorrect boundary conditions (e.g., no-flow or closed-basin boundaries when they are not appropriate) or errors that result because the conceptual model is incorrect (e.g., a gaining stream is assumed when it is really a losing stream, or assuming a porous media aquifer when it is really a fractured rock aquifer, etc).

Multi-model averaging – is a means to address the prediction uncertainty associated with model error, rather than the uncertainty associated with unknown or uncertain input parameters for a fixed conceptual and mathematical model. In this approach, different models agreed upon by stakeholders can be used to obtain predictions, and the results of these different models can be averaged to obtain a more representative predicted condition. The different models can also be used to place bounds on the model predictions.

Model verification and validation – terms that are often interchanged, confused, and misused in the hydrogeology literature and in the environmental consulting and regulatory arena. The researchers adopted the definitions similar to those that are universally accepted by most engineering textbooks.

- Model ***verification*** is the process of ensuring that the model correctly reproduces the mathematics and algorithms of the model. This would include verifying the correctness of any numerical methods associated with the model, checking the computer programming used to create the model software, and determining the veracity of any graphical user interfaces that are used with the model. Most commercially available models have been verified by comparing the results to other validated models.
- Model ***validation*** is the process of determining the degree to which a model is an accurate representation of the processes that are being modeled. Most textbooks recommend history matching (i.e., model calibration) as the preferred model-validation technique. Some textbooks consider validation to be different than calibration, and define the word as a test of whether a calibrated model can adequately simulate data that was not used during the calibration process. Expert groundwater modelers are increasingly recommending that the word validation is misleading and should not be used. That is, one might expect that a "validated" model can be used for decision making indefinitely. The researchers agree with Bredehoeft and Konikow (1993) that a model can never be validated, it can only be invalidated. However, the performance of any model must be continuously evaluated (much like a car requires consistent maintenance). The user's guide will adhere to this latter philosophy, and thus will often use the general term "model performance evaluation" (described below) instead of the "validation."

Post-modeling audits – the process of checking that the model is continuing to serve the intended purpose beyond the period of the first calibration and use for decision making.

Model-performance evaluation – an assessment that determines fallibilities. Often the stresses in the watershed hydrologic system that occur after calibration are significantly different than those that occurred during the calibration period. In other words, the time period used for calibration was not long enough to capture all the important hydrologic and pollutant-transport processes that are likely to occur during the period for which the model use is relevant (i.e., the time period for which the planning decision is applicable). If the model is used for long-term predictions, then the modeling process needs to be continually evaluated beyond the initial calibration to ensure that the model predictions continue to "hold water." This includes the initial calibration, the post-audit, and making changes as a result of the post-audit. This might involve adjusting the calibration parameters as new data are obtained, or could be as drastic as changing the entire conceptual model and starting over if the existing model is not doing the intended job.

4.6 What Does it Take to Make a Good Model?

Many things must be considered regarding development of a useful mathematical model to aid in decision making. Listed below are some that the researchers consider to be important. The list is not a ranking, because the researchers believe all these factors are critically important to development of a "good model."

1. Understand the purpose of a model. Models are used to aid decision makers. The first decision that must be made is defining the model's purpose. The level of funding and commitment to model development will depend on the eventual use of the model. Models that are to be used as screening tools to move forward to the next step of the decision-making process do not require as much commitment as models that will be used to make predictions. Various commitment levels are required to construct models that are used to make decisions or predictions regarding, human health, environmental health, costs associated with land-use options, or aesthetics. More details on this topic are provided in Chapters 5.0 and 7.0.

2. Produce a good conceptual model. The user should develop an appropriate conceptual model (defined earlier) for the natural, regulatory, and economic system for which the model is to be constructed. All future choices regarding model selection and development are based on this conceptual model, including the type of model (e.g., groundwater model, surface-water model, watershed model, and vadose zone model), the appropriate level of complexity (numerical distributed model, GIS screening model, or simple mass-balance screening model), amount of additional data collection, and model-development budget. More details on this topic are in Chapters 5.0 and 6.0.

3. Consider the processes when selecting the appropriate model.. Models that are designed to simulate surface-water processes are not likely to be appropriate if the primary goal is to assess groundwater contamination. Similarly, one should not choose models that cannot appropriately simulate the processes of interest (e.g., the nature and composition of the OWS source, or processes associated with a particular pollutant of interest). While no commercially available model is likely to be appropriate for all possible situations that may occur in a watershed, the model should be able to simulate the priority problems that are identified by the stakeholders. More details on this topic are included in Chapters 5.0 and 6.0.

4. Model complexity should be commensurate with the problem. Some problems may be sufficiently addressed using a relatively simple screening model. Other problems may require a more complex model to enable the resulting decisions to be rigorously justified. However, if a complex model is used for a problem that could have been addressed with a simpler model, then the use of resources and money has not been optimized. For most problems, it is wise to start simple, and progress to more complex models as needed.

5. The model should be unique. Complex models with many input parameters should be evaluated to determine if the model result is truly unique. That is, some input parameters are correlated such that several different combinations of values for the different input parameters may result in the same model output result. However, only one of these combinations could be theoretically valid. Models that are not unique are not likely to provide good results beyond the time period for which the model was calibrated or evaluated. Model uniqueness is discussed in more detail in Chapter 9.0.

6. The uncertainty associated with the model output should be explained. No model provides an exact answer. There is uncertainty associated with the input data used in the model, and indeed with the model itself, that result in uncertainty in the model's output data. This uncertainty should be considered in the decision making process. Uncertainty analysis is discussed in more detail in Chapters 9.0 and 10.0.

7. Appropriate data must be obtained to enable an evaluation of the model's performance. Complex models generally require more data. If the decision to use a complex model is based on the consequences and risks associated with the decision, then a commitment should be made to collect the appropriate data to use that model. In some cases this is data used for model calibration. However, it may also be data collected to understand the variability of the input so that the uncertainty in the model output can be evaluated. Complex models in particular are susceptible to non-uniqueness (described above) if insufficient data are used for calibration. Models should *not* be selected based on the amount of data available. If this is done because of time or cost considerations, then stakeholders should factor this into their decision-making process.

8. The model uses a broad range of information to constrain the conceptual and mathematical model and the model results. It is useful to consider "soft" data when evaluating model performance. Examples include important knowledge about the natural system that cannot be quantified with traditional recorded data.

9. An appropriate level of support must be dedicated to model development after considering the issues above. Stakeholders must ask themselves what health, financial and legal risks and costs are likely to be associated with a poor or poorly justified decision. The amount of resources dedicated to quantitative analysis (i.e., modeling) to aid in the decision-making process should be commensurate with these risks and costs.

10. All major stakeholders should "buy in" to the modeling process before it starts, including all aspects of the model-development process listed above.

4.7 General Modeling Approach – Parsimony

"A model should be as simple as possible, but no simpler." (Attributed to Albert Einstein)

"For every complex problem, there is a simple, and wrong, answer." (Attributed to Albert Einstein)

Sometimes it is necessary to use complex modeling to accomplish the stakeholder's goals. However, it is rarely wise to start the analytical process with a complex model. In all cases, simpler screening models that make conservative assumptions and/or assume worst-case conditions should be applied first. If results of these screening models suggest a problem may exist, then models of increasing complexity can be used as the situation warrants. In addition, even when complex models are chosen, the researchers agree with Hill (1998) that it is useful to start simple and increase complexity in the model as necessary.

(Much of the remaining discussion in this section is adapted from Hill (1998), whose focus is primarily on groundwater models, but is applicable to watershed-scale models. Hill points out that a model should remain as simple as possible, yet still account for the relevant dominant processes that are justifiable based on the observed data or characteristics of the natural system.)

Watershed-scale models typically consider a large number of hydrologic and pollutant transport processes related to climate, overland sediment transport, infiltration into and through the vadose zone, groundwater hydrogeology and transport, and stream hydrology and transport.

Thus, one should incorporate the processes that are likely to be dominant, and add complexity as it is needed, and as one better understands the system (Hill, 1998). Ideally, one should evaluate the model performance with measured data or other information known about the system to test whether the added complexity is warranted. This approach may appear at first to be less efficient, but the mild investment in parsimony is almost certain to result in a better understanding of the system, and thus in a better model. In addition, this approach may also save money. If added complexity does not improve the model, then the added complexity can only serve to increase the cost of the model and limit its credibility with stakeholders.

After each step below, decision makers should evaluate whether moving to a more complex model is warranted based on available data and stakeholder input. In addition, the decision to move forward may require collection of additional data.

1. Start with a simple screening model that includes the most important hydraulic compartment of interest (e.g., groundwater, surface water, a lake, etc).
2. If warranted, proceed to a more complex, but relatively simple model (e.g., lumped parameter model, analytical solution, or GIS-based model) that considers the hydrologic compartment of interest.
3. If warranted, proceed to a spatially distributed model that accounts for the most important hydrologic compartment of interest. Incorporate the most dominant features of the hydrologic system.
4. If needed, add features to the model that honor the hydrologic system.
5. Add additional hydrologic compartments as warranted, which may require use of an alternate model. Incorporate the most dominant features of the hydrologic system.

In summary, models should be as complex as necessary, but as simple as possible.

Adding complexity should be done carefully and sequentially. Adding complexity that existing data does not support will result in a model that likely cannot be uniquely calibrated or cannot be rigorously defended.

4.8 Relationship between Modeling, Risk Assessment, and Decision Making

Models are useful to aid in making decisions regarding environmental health, land-use planning, and development. There is some risk associated with making any policy decision: the decision may fail to prevent adverse impacts to environmental or human health. The main risk arises from a poor understanding of the possible environmental outcomes that can result from implementing a decision. Models cannot eliminate this risk! Rather, models can help decisionmakers understand the uncertainty associated with implementation of a particular decision. , Stakeholders can then better evaluate the risk associated with a particular decision.

One of the best uses of models is to conduct uncertainty analyses (see Chapters 9.0 and 10.0 for more technical details on this topic). Regardless of the uncertainty analysis approach, the goal is to better understand the uncertainty in the model prediction, which is usually based on the uncertainty in the model input, or equivalently, the uncertainty in the data used to characterize and describe the hydrologic system.

A model should rarely be used to produce a single result, and a decision should rarely be based on a single model result. This implies a certainty associated with a particular land-use implementation or decision that does not exist. Rather, the honest approach uses a model to develop a range of probable outcomes, which provides more realistic information upon which to base a decision. Often, such an approach frustrates the decision makers because there is a presumption that the model will come up with one answer that will enable a unanimous decision. This is certainly not a realistic expectation.

Why, then, use a model? Without a model, it would be nearly impossible to evaluate the potential outcomes associated with a decision, and to obtain a measure of the relative likelihood of a particular outcome. In addition, without using a model that can incorporate relevant processes, overly conservative decisions are often made.

A typical example is related to assessments of potential nitrogen impacts from OWS. Most assessments assume that nitrate is completely conservative; that is, that nitrate is not lost due to the process of denitrification. It is known that denitrification occurs in nearly all soils, albeit to various extents. Nonetheless, denitrification is difficult to estimate or measure, and is thus usually omitted. However, if a rigorous effort is made to understand the natural variability in the denitrification process, and this information is combined with a model that can account for the process of denitrification, then a range of potential outcomes can be developed (e.g., nitrogen concentrations in groundwater relative to a regulatory limit). The range of outcomes can be very large, or can be surprisingly small, depending on the specific site and scenario being modeled. Decision makers can evaluate the various model outcomes relative to the input assumptions required to develop each outcome, and make a decision based on their aversion to, or willingness to accept, risk. Willingness to accept risk almost always depends on the consequences or benefits associated with the decision that poses the risks.

In another example, Heatwole and McCray (2007) conducted a modeling effort to help decide whether OWS could be used for a large development where very sparse data were available. (This case study is also described in Section 4.10 of this chapter.) Initially two screening models typically used for nitrogen assessments in aquifers (that neglected denitrification), predicted that concentrations in the aquifer would far exceed the regulated maximum concentration limit (MCL). Next, a model was used that accounted for the denitrification rate in the top layers of the soil (and some other important processes). A larger denitrification rate results in more nitrates being removed during transport in the subsurface.

The model was developed using cumulative frequency diagrams (CFDs), which illustrate the frequency of a particular value, in this case the frequency of transport-model input parameter values occurring among all those reported in the literature. More than 100 values from different soils and sites were obtained. Figure 4-1 shows a CFD for denitrification rate. (The modeling study associated with this example is described in more detail in section 4.10 and in Chapter 11.0.) The model predicted that the MCL would not be exceeded as long as the true (but unknown) denitrification rate exceeded a relatively small value; a value that was smaller than 85% of all values reported in the literature (i.e., the 15th percentile value). For any assumed value larger than this 23rd percentile value, an exceedence of the MCL was not predicted. This result enabled the stakeholders to discuss the relative risks associated with the decision (i.e., whether or not to allow the development to use OWS for wastewater treatment).

Most stakeholders agreed that using the small 15th percentile value for the denitrification rate was a conservative (but not overly conservative) assumption, and were willing to accept the risk that the actual value might be smaller than assumed, and could thus result in actual groundwater contamination. Therefore, they decided to permit OWS for the development. Of course, a few stakeholders exhibited a high aversion to risk, and thought that even this relatively small risk should not be accepted because of the importance of ensuring water quality in the shallow aquifer. To help alleviate the concerns of those who had a high aversion to risk, installation of some groundwater monitoring wells were recommended to enable action to mitigate off-site migration of any contamination that did actually occur.

Suppose a similar modeling effort at another proposed development site had determined that a much higher value of the denitrification rate would predict an exceedence of the MCL (e.g., a 80th percentile value, whereby 80% of all rates reported in the literature were less than this rate). In that case, it would be likely that nearly all of the decision makers would *not* be willing to accept the associated risks. Perhaps OWS would not be permitted, and a wastewater treatment plant would be built.

For another case, suppose a 35th percentile value would have resulted in an exceedence of an MCL. One could interpret this as suggesting that there is a 35% probability that OWS would result in violating a regulatory limit. This scenario might split stakeholders "down the middle," where one group would be willing to accept that risk to enable development, and the other group would not be willing to accept that risk of aquifer contamination.

These examples illustrate that decisions are based on the stakeholders' willingness to accept the risks relative to the perceived benefits associated with a decision. In the above scenario, it would have been very difficult to quantify the risks in the face of sparse data, and make an informed decision based on these risks, without using a relatively complex mathematical model that incorporated the important natural processes. If typical screening models were used, OWS likely would not have been permitted because they all predicted concentrations several times higher than the MCL. After results from an appropriate model were considered, it seemed clear that the risks associated with potential groundwater contamination from the proposed OWS were acceptable.

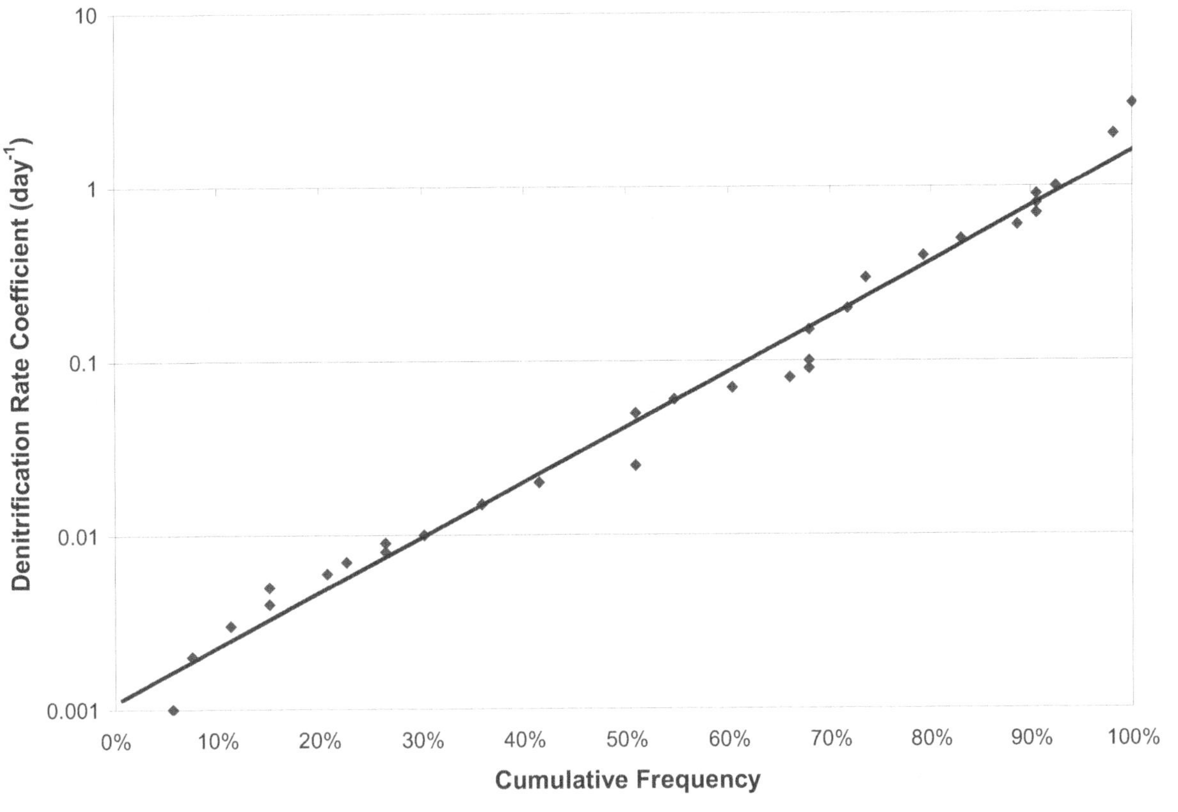

Figure 4-1. Cumulative Frequency Diagram (CFD) for Denitrification Rate.
From Heatwole and McCray, 2007. Reprinted with permission from Elsevier.

In the case study above, insufficient data were available to calibrate a model because development was in the planning stages. Thus, a decision was made to collect sparse hydrological data useful for model input because of time constraints associated with the timeline set for the decision making process. Of course, the most prudent approach would be to collect data on denitrification rates at the site so that the potential range of denitrification rates could be better determined. This would reduce the uncertainty associated with the decision.

4.9 Relationship between Prediction and Management Scenario Analysis

Watershed models can help evaluate the impact of various proposed or hypothesized land-use and water-use decisions. Many different scenarios can be tested and the resulting impacts on water quantity or quality can be evaluated relative to each other.

A few examples of possible scenarios:

- OWS versus centralized sewers for a planned development
- OWS versus decentralized cluster systems for a planned development
- Residential development with OWS versus other potential uses (e.g., agricultural land, forest, commercial development)
- Impact of OWS versus centralized systems on water quality and water quantity in a watershed

- Evaluation of the optimal locations within a watershed for development to minimize water-quality impacts
- Pre-development community-scale assessment with sparse data.

If a model is calibrated using data from a watershed, then users have much more confidence that the model results are accurate enough that planning decisions can be based on the results. However, it is not reasonable to expect that even rigorously calibrated models can reliably make predictions over the long term (discussed below). Nonetheless, models can be extremely useful when used to evaluate the *relative* impacts of several different future management scenarios. Of course, one should be careful not to term the results for a particular scenario as a prediction. Rather, the results should be evaluated in terms of assessing which scenario yields the most or least adverse impact. It may be prudent to commission a more detailed modeling study on one or two of the most amenable future land-use scenarios to obtain a more accurate view of the potential impacts.

Why should modeling results, used to evaluate different scenarios, not be viewed as predictions? Ideally, watershed models can be rigorously calibrated using long-term hydrologic and pollutant data (more than 20 years) collected from several locations in the watershed. Such a model could presumably be used reliably for predictions 20 years into the future assuming hydrologic and land-use conditions did not change considerably during that time. However, such data are rarely available in any watershed. This is especially true for watersheds or portions of watersheds where OWS development is proposed because the locations are generally sparsely populated, current impacts are negligible, and there is no reason to collect pollutant data. Even in urban watersheds, where the hydrologic data record can be very good; data for pollutants typically associated with OWS is rare. In addition, land-use changes that can have a serious impact on the hydrology and pollutant loading in a watershed are almost certain to occur given that the watershed is already of interest related to OWS and therefore development. Thus, it is not reasonable to expect that even rigorously calibrated models can be used to reliably make accurate predictions over the long term. Rather, models are calibrated so that the user and stakeholders have confidence that the model can adequately represent the watershed under current conditions. Then, the model can be more confidently used to evaluate future decisions related to land or water use, even if the model cannot be calibrated under those conditions, provided that the model is designed to simulate those future land- or water-use changes.

Finally, when future scenarios are projected, it is important that the model be capable of simulating the specific projected change, and that the hydrologic and pollutant impacts of those changes be carefully considered and incorporated into the model. For example, in a watershed where phosphorus (P) loading to surface water is of concern, implementing OWS in a proposed development is also of concern. However, depending on the soils in the watershed, the P loading from sediment released during construction activities and land-surface changes related to development could potentially release much more phosphorus to surface water than could be added from OWS. Therefore, one should use a model (or models) that can simulate introduction of P into the stream via overland sediment transport, as well as a model that can be used to evaluate P transport and transformations through the subsurface and eventually into the streams.

Chapter 11.0 provides several different modeling case studies, which illustrate some common or interesting scenario evaluations related to OWS.

4.10 Description of a Modeling Case Study

The case study described in this section is presented in more detail by Heatwole and McCray (2007). This paper describes a modeling effort to help local environmental regulators and county officials decide if conventional OWS is appropriate in a proposed 1100-home development. The environmental/ hydrologic systems of interest were: the nearby river that is a source of drinking water for several cities and two states; an alluvial aquifer associated with the river; a shallow aquifer below the development that was used by current inhabitants; and a deeper semi-confined aquifer that is used by many towns and cities as the primary drinking-water source. The first step was to develop a *conceptual model* upon which to base the mathematical model. An illustration of the conceptual model is shown in Figure 4-2.

The data showed that that the deeper aquifer was well confined below the proposed development by a shale unit, and thus the potential for contamination would be negligible. Data analysis also demonstrated that the shallow aquifer was likely not hydraulically connected to the alluvial aquifer around the adjacent river. There was some uncertainty related to this assessment. However, the water in the adjacent river was already relatively high in nitrates from wastewater-treatment plant discharges and upstream agriculture. The river and the alluvial aquifer had been shown previously to be "well mixed," and the nitrogen concentrations in this alluvial aquifer were also relatively high.

The primary concern remaining was the heterogeneous shallow aquifer below the proposed development, which was used by several existing homes in the area. This aquifer was sparsely used, and some suggested that any potential contamination could be corrected for the relatively few homes by providing water treatment systems. However, both the county and state departments of health and environment believed that the sparsely used shallow aquifer should be protected because of the likelihood of water-scarcity in the future. In other words, the *perceived risk* of shallow-aquifer contamination from OWS in the development was deemed significant enough that a resulting *decision* was made to evaluate the potential for nitrogen contamination. If it was determined that the shallow aquifer was likely to be contaminated above a specified limit, then additional evaluative steps might be taken to assess whether this shallow sedimentary aquifer was connected to the alluvial aquifer associated with the river, and whether this additional nitrate would have an impact on this alluvial system. This would be the *simplest conceptual model possible* that still honored the information and data that were evaluated.

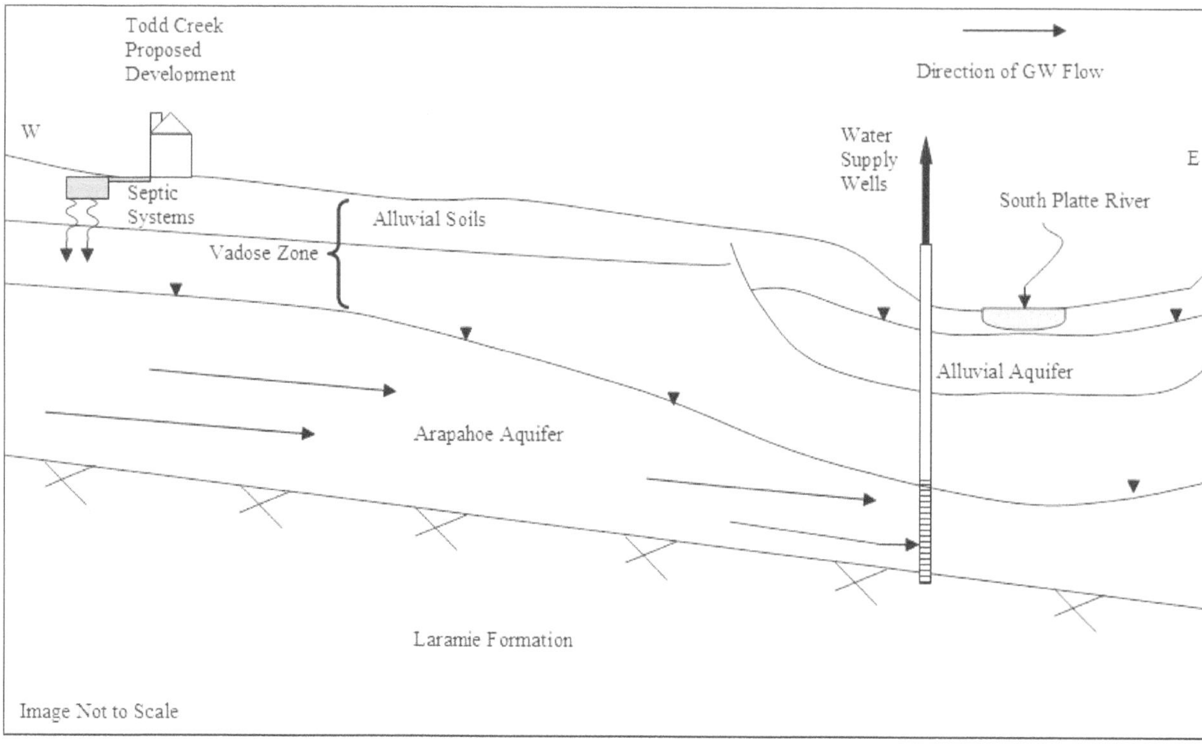

Figure 4-2. Conceptual Model for Scenario Discussed in Section 4.10.
From Heatwole and McCray, 2007. Reprinted with permission from Elsevier.

The next step was to apply a *simple mathematical screening model*, implemented with a spreadsheet, to determine whether nitrate contamination was possible. The selected model was similar to the well-known Wehrmann (1984) model and assumed that all the nitrogen from the OWS would reach the aquifer, and then would mix in the aquifer over a specified depth.

The groundwater flow rate, the OWS input rate, and the mixing depth were inputs to the model.

The model generated several outcomes, using regional data to obtain the groundwater flow rates, population information to estimate the OWS input, published median values for nitrogen in STE from McCray et al. (2005), and several mixing depths that were deemed reasonable based on known contaminant plumes that originate from near the land surface,. In all cases, the screening model showed aquifer nitrate-N concentrations (i.e., reported as N) in excess of 30 mg/L (the MCL for nitrate-N is 10 mg/L). However, it was known *a priori* that this model was conservative (no denitrification, and no contaminant dispersion in the vadose zone or in the aquifer), and would therefore over-predict the nitrate concentration.

The researchers decided to use a somewhat more complex model that could account for dispersion, and that could be implemented into a spreadsheet. This model was first proposed for contaminant transport by Galya (1987). Dispersion is a mixing (and dilution) process caused primarily by heterogeneities in the subsurface. Conservative estimates of dispersion were accounted for in this model. This model also showed long-term aquifer nitrate concentrations considerably higher than the MCL, but that were considerably smaller than that predicted by the simpler model.

After evaluating the results of these screening models, stakeholders agreed it was worth the effort and expense to consider a *more complex conceptual model* that considered actual dispersion that might be occurring at the site, as well as denitrification, which is known to occur in nearly all soils to varying degrees. The conceptual model considered that denitrification would likely be strongest in the shallow soils immediately below the OWS effluent (which were more likely to be anaerobic, and higher in organic matter, both required conditions for denitrification).

The first step was to conduct a *model sensitivity analysis*, with results summarized in Figure 4-3. The figure shows that the model output (nitrogen concentrations in the shallow groundwater) were most sensitive to the uncertainties in denitrification rate, the OWS flow rate, and the nitrogen concentrations in the STE (Heatwole and McCray, 2007). Given the large number of homes in the area, it was felt that using median rates for flow and nitrogen concentrations were reasonable. In addition, many measurements have been reported in the literature for these parameters, and the variability and uncertainty in these measured rates were not nearly as large as for the denitrification rates.

Implementation of a mathematical model for this conceptual model required collection of additional data that would enable application of a more complex mathematical model. Two test holes were drilled to depths of about 20 feet at the site to better understand the subsurface hydraulic properties. Organic matter content was also measured to ensure that denitrification was feasible. However, denitrification rates were not measured due to the complexity of the measurement technique for field soils, and because of budget constraints. Thus, uncertainty in denitrification rates was considered in model application (recall Figure 4-1).

The model results are summarized in Figure 4-4. The model that used soil-hydraulic data based on the boreholes drilled at the site predicted that the MCL would not be exceeded as long as the true (but unknown) denitrification rate exceeded a relatively small value; a value that was smaller than 85% of all values reported in the literature (i.e., the 15th percentile value). For any assumed value larger than this 23rd percentile value, an exceedance of the MCL was not predicted. If average soil hydraulic properties based only on the general soil-survey description were used (which classified the soil as a Platner loam in general), then the MCL was not exceeded if a 22nd percentile value for the denitrification rate were used. This simulation was conducted to enable a better understanding of how site-specific hydraulic data might show different results than site-averaged data. It is possible that locations different than the test hole locations might show results more closely associated with the results for the Platner loam.

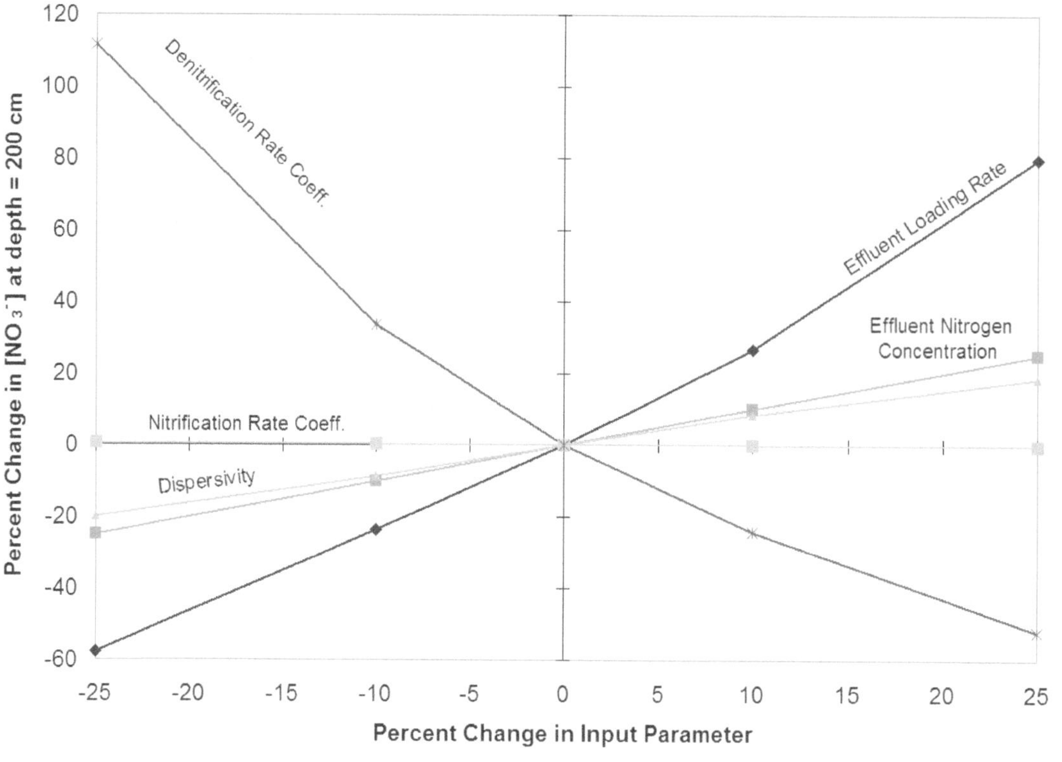

Figure 4-3. Results of Sensitivity Analysis for Scenario Discussed in Section 4.10.
From Heatwole and McCray, 2007. Reprinted with permission from Elsevier.

The modeling effort enabled the stakeholders to engage in a discussion about the relative risks associated with the decision (i.e., whether or not to allow the development to use OWS for wastewater treatment). For example, most stakeholders agreed that using either the small 15th percentile value or the 22nd percentile value for the denitrification rate was a conservative (but not overly conservative) assumption, and were willing to accept the risk that the actual value might be smaller than assumed (i.e., less denitrification than assumed), and would thus more likely result in groundwater contamination. Therefore, a decision was made to permit OWS for the development.

Of course, a few stakeholders expressed a high aversion to risk, and thought that even this relatively small risk should not be accepted because of the importance of ensuring water quality in the shallow aquifer. To help alleviate the concerns of those who had a high aversion to risk, installation of some groundwater monitoring wells were recommended with quarterly monitoring to enable stakeholders to take action to mitigate off-site migration of any contamination that did actually occur below the development.

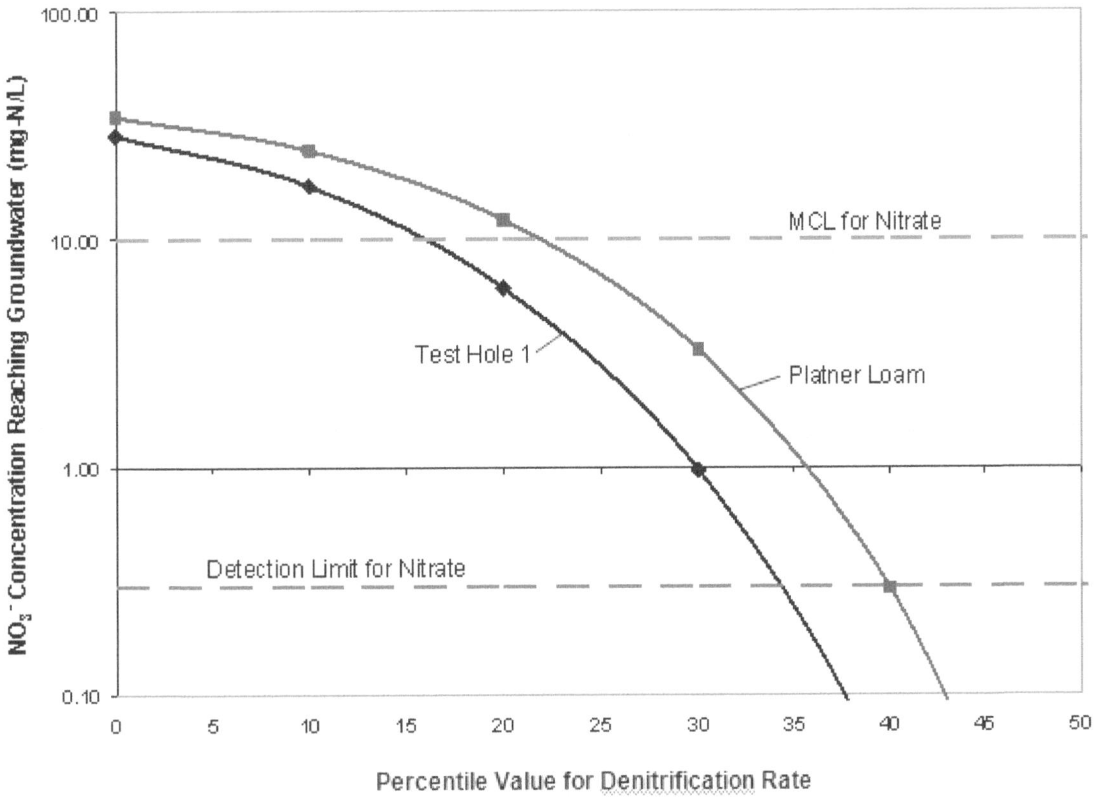

Figure 4-4. Model Results for Two Soils Accounting for Uncertain Denitrification Rates
From Heatwole and McCray, 2007. Reprinted with permission from Elsevier.

CHAPTER 5.0

MODELING APPROACHES AND MODEL TYPES

5.1 Mathematical Representation of Hydrologic Systems

5.1.1 Mass-Balance Equations

Any hydrologic system can be represented by applying the law of conservation of mass. When constructing a mass-balance model the mass (e.g., of water or nutrients) entering a system must either leave the system or accumulate within the system. All mass-balance equations take the form of:

$$Input = Output + Accumulation$$

The boundaries of the system (e.g., watershed) must be well-defined so that inputs and outputs crossing the system boundary can be quantified or estimated. Mass-balance equations can be applied to hydrologic systems by using measured data of inputs/outputs, approximating unmeasured parameters using empirical or analytical equations, and making simplifying assumptions. Simplifying assumptions may include: 1) which inputs and outputs are negligible relative to other system components, 2) steady-state, which makes the accumulation term zero.

Example: Watershed-scale modeling of OWS might include a mass-balance equation to estimate mass of nitrogen released at the mouth of the watershed.

5.1.2 Empirical Equations

An empirical equation is one based solely on observation rather than theory, and forms a relationship between a desired, difficult to measure parameter and a commonly measured parameter. An empirical relationship requires only confirmatory data irrespective of theoretical basis.

Example: Watershed-scale modeling of OWS might include the use of the empirically based USDA Soil Conservation Service (SCS, 1972) curve number (CN) method for rainfall-runoff modeling.

5.1.3 Analytical Equations

Analytical solutions are possible for mathematical models that have a closed form solution. In other words, changes in a system can be expressed as a mathematical function with one exact solution. These solutions often assume a steady-state or only apply to models with a single time-step.

Example: Watershed-scale modeling of OWS might include the use of an analytical solution to estimate the denitrification rate of an OWS.

5.1.4 Numerical Equations

Numerical solutions are required when the governing equations are ordinary differential equations (ODE) or partial differential equations (PDE). Numerical solutions do not seek exact answers because exact answers are impossible to obtain in practice. Numerical approximations are achieved using a computer code that solves a set of algebraic equations that approximate the governing equation, boundary conditions, and initial conditions of the mathematical model. These solutions are often necessary when multiple time-steps are required to simulate a process in the watershed. The approximate solution to one ODE or PDE becomes input to another ODE or PDE in a time-series analysis.

Example: Watershed-scale modeling of OWS might include the use of a numerical solution to solve the groundwater flow equation (GWFE) or advection-dispersion equation (ADE).

5.2 Types of Hydrologic Models

If a model is necessary to understand current or anticipated nutrient loads from OWS at the watershed-scale, then the selection of the model type becomes critical to planning and management of the watershed. The type of model should be selected by first clearly defining the nutrient and OWS-related question or concern, then selecting the most appropriate type of model. The selected model type can be used to cost-effectively answer the nutrient and OWS-related question within an acceptable level of risk.

A screening model is any model used to evaluate a system using highly simplified relationships between system inputs and outputs and requiring minimal data. When completing watershed-scale modeling of decentralized wastewater systems, screening models can be used to identify potentially sensitive areas, to highlight areas where more data are needed, or to quickly test alternative conceptual models. Screening models can be used to determine whether impacts from OWS are likely under worst-case scenarios (e.g., no mixing or attenuation). This type of model can be used to determine whether more resources should be devoted to data gathering or toward development of more complex models or quantitative tools. Consequently, screening models provide cost-effective evaluations on strategies before more expensive physically-based models are employed for a specific site.

Screening models and tools that implement simplifying and conservative assumptions should be used to first evaluate if OWS impact is a potential concern within the limits of uncertainty associated with the model results. If OWS impact is shown to be a reasonable concern, then models with increasing complexity and implementation cost can be used in a sequential manner until the decision makers are comfortable with the model predictions, including the model uncertainty, relative to the risk associated with the decisions.

The types of models listed below are arranged, generally, in order from least complex to most complex. Increased complexity normally translates to higher costs for gathering of necessary data, implementation and calibration of the model. In all cases, it is suggested that the watershed-scale OWS question be analyzed with the simplest model type and progress to more complex model types as necessary. Mass-balance screening models usually produce a conservative estimate of nutrient loads, which allows the user to either be: 1) satisfied that there is not a problem (due to conservative estimate) and end the modeling process; or 2) unsure of severity of problem so continue refining the "answer" by implementing more complex model types.

5.2.1 Mass-Balance Screening Models

Mass-balance screening models are normally implemented in a spreadsheet and are highly simplified estimates of water and nutrient balances.

Example models include: HPS, OSF

- HPS: The Horizontal Plane Source model is a transient, three-dimensional analytical model, capable of simulating a horizontal-dispersive movement in a homogenous, isotropic aquifer (Heatwole and McCray, 2007). This model is less complex and requires fewer data inputs than other existing modeling approaches that have been proposed, making it a suitable model for local and county environmental agencies in evaluating OWS groundwater impacts at a development scale.
- OSF: The On-Site and Fertilizer model was applied for land management purposes in Martha's Vineyard, Massachusetts (Gaines, 1986) and is the simplest screening model. OSF assumes that N inputs in the watershed are from homes, and simply adds an annual load of septic N (6.8 kg N a^{-1}) and lawn fertilizer N (4.8 kg N a^{-1}) per household. Atmospheric N loads are assumed to be taken up within the watershed (Valiela et al., 2002). The number of homes in the watershed is the only data requirement, so it will provide a worst-case scenario for N loading to the mouth of the watershed.

5.2.2 GIS-Based Screening Models

GIS-based screening models are implemented either completely within a GIS or require the use of GIS for processing of geographically-referenced data (e.g., land-use, soil property and precipitation) as primary model inputs. A GIS-based screening model is a model that combines a simple conceptual model and analytical equations that can be solved using a GIS. It generally captures the regional spatial characteristics of a system without including the detailed processes involved or data density that would be necessary for site-specific delineations. When integrated within GIS, computations become powerful and time-efficient because of advantages in data storage, manipulation, analysis, and visualization of spatial data.

Example models include: MANAGE, NLM, and PLSM

- MANAGE: The Method for Assessment, Nutrient-loading, and Geographic Evaluation of non-point pollution model (Kellogg et al., 1996) is a spreadsheet (i.e., Microsoft Excel™) based model that uses input derived from spatial analysis in GIS. This uncoupled (Excel to GIS) watershed assessment tool can evaluate pollution risks of land use and landscape features. MANAGE functions include the identification of areas where natural features and high intensity land uses together increase the risk of nutrient runoff to aquifers and surface waters, comparing the effects of existing and future land use patterns on water resources, and evaluating the effectiveness of storm and wastewater management practices for reducing pollution risk. The MANAGE model consists of two components for assessing nitrogen contributions to groundwater from OWS, surface water and groundwater. The surface water component of MANAGE uses published export coefficients to estimate N and P loads from 21 land use types. The groundwater component assumes that 80% of the N in OWS enters the aquifer without estimating any losses of NO_3. Information on how to obtain the model is given in Appendix C.
- NLM: The Nitrogen Loading Model (Valiela et al., 1997; Valiela et al., 2002) predicts total dissolved N loads to shallow estuaries from rural suburban watersheds where groundwater

flow is the dominant transport vehicle. The NLM is complex enough to represent the nature of the systems and sufficiently adapted to local conditions to produce accurate predictions, while simple enough to be applied to different types of situations. NLM uses values for per capita contributions to estimate N inputs from DWTS that are derived from GIS data representing the total number of residences. NLM also accounts for losses of nitrogen within septic systems (by denitrification, volatilization of ammonia, or by adsorption of ammonium) as well as within leaching fields. Further information can be obtained from sources provided in Appendix C.

- PLSM: The St. Johns River Water Management District created the Pollutant Load Screening model to cope with Florida's ever-increasing population and the resulting pressure on the water quality of lakes and rivers. The PLSM is a GIS-based watershed model for estimating runoff and annual pollutant loads (Adamus and Bergman, 1995). Runoff for each land use and soils combination within the study area is first determined by multiplying average annual rainfall, a runoff coefficient that depends on soil and land use type, and the area of the basin under study. The annual pollutant load is then determined by multiplying runoff by a runoff pollutant concentration coefficient that depends on the type of land use. Further information can be obtained from sources provided in Appendix C.

5.2.3 Surface-Water Models

Surface-water models focus on fluxes between the atmosphere and ground-surface and changes in storage within rivers, streams and other surface waterbodies. A surface-water model normally combines empirical models and/or numerical models to simulate flow rates and nutrient concentrations over various time-scales.

Example models include: ANSWERS-Continuous, HSPF

- ANSWERS-Continuous: The Areal Nonpoint Source Watershed Environment Response Simulation (Beasley et al., 1980) was modified to include a continuous simulation mode (Bouraoui and Dillaha, 1996). ANSWERS-Continuous operates on a daily water balance with nutrient losses through uptake, runoff, and sediment. Does not include a direct mechanism for releasing septic system effluent into the model.
- HSPF: The Hydrological Simulation Program – Fortran model represents many of the processes of interception, infiltration, evapotranspiration, snowmelt, surface runoff, interflow, groundwater loss, groundwater recharge, and base flow using empirical equations (Donigan et al., 1984). Does not include a direct mechanism for releasing septic system effluent into the model.

5.2.4 Vadose-Zone Models

Vadose zone models are built around the unsaturated zone and consider fluxes between the overlying ground surface and the underlying saturated zone.

Example model is: HYDRUS

- HYDRUS: Numerically solves the Richards equation for saturated-unsaturated water flow and convection-dispersion equations for heat and solute transport (Simunek et al., 1999). The transport equations include provisions for reactions between the solid and liquid phases along with first-order degradation reactions.

5.2.5 Groundwater Models

Groundwater models focus on changes in storage and fluxes within the saturated zone. Normally they are a combination of empirical models and/or numerical models that can predict hydraulic heads and nutrient concentrations over various time-scales.

Example models include: FEFLOW, MODFLOW, SUTRA

- FEFLOW: The Finite Element subsurface FLOW and transport modeling system was designed for simulation of unsaturated and saturated groundwater flow. FEFLOW allows for full 3-D finite-element modeling of contaminant transport, including reaction and sorption kinetics.
- MODFLOW: The MODular three-dimensional groundwater FLOW model numerically solves the groundwater flow equation for saturated porous medium by using a finite-difference method (Hill, 1992). MODFLOW can be used in conjunction with compatible modules such as MT3D (Zheng and Wang, 1999; Zheng, 2993) or RT3D (Clement 1997, Clement et al., 1998) to simulate contaminant transport and reaction. It can also be used in conjunction with other models to allow for modeling of flow and transport coupled between surface water, vadose zone and saturated groundwater zones.
- SUTRA: The Saturated Unsaturated Transport model is a numerical code that accounts for variable density fluid flow in saturated or unsaturated conditions with solute or energy transport (Voss, 1984). SUTRA solves Darcy's general flow equation using permeability (k), instead of hydraulic conductivity (K), and thus can account for variable fluid density in the numerical flow equation. SUTRA employs a two-dimensional, finite element method with a finite difference approximation in time to simulate variable-density groundwater flow and solute transport using pressure inputs and the ADE.

5.2.6 Integrated Watershed Models

Integrated watershed models include surface water, vadose zone, and groundwater flow modules; allowing for temporal calculations of flux and storage in each component of the hydrologic system.

Example models include: GWLF, MIKESHE, SWAT, WARMF

- GWLF: The Generalized Watershed Loading Function model is a lumped parameter watershed-scale model that uses hydrology, land cover, soils, topography, weather, pollutant discharges and other environmental characteristics to assess non-point source pollution and sediment and nutrient loading (N and P) from a given watershed (Haith and Shoemaker, 1987). GWLF is a continuous simulation model, using daily time steps for weather data and water balance calculations. Monthly calculations are used for sediment and nutrient loads based on the daily water balance accumulated to monthly values. GWLF is implemented through an ArcView GIS interface and referred to as ArcView GWLF (AVGWLF) (Evans et al., 2006).
- MIKE SHE: based on the Systeme Hydrologique Europeen model is comprised of five sub-modules: evapotranspiration (ET), unsaturated zone flow (UZ), saturated zone flow (SZ), overland and channel flow (OC) and irrigation (IR). The model is largely physically based and runs as a finite-difference numerical model. Does not include a specific mechanism for releasing septic system effluent in the model.

- SWAT: The Soil and Water Assessment Tool model is a watershed model most often used in TMDL assessments. SWAT has not been modified to directly simulate OWS, but can be used to evaluate the effect of OWS by making certain assumptions. SWAT is known for its ability to simulate many hydrologic processes such as surface runoff, lateral flow, groundwater flow, sediment loading, N, P, and pesticides. SWAT operates on a daily time step and requires, at a minimum, the following input data: climatic, digital elevation, land use-land cover, and soils data. SWAT was developed to predict the impact of land management practices on water, sediment, and agricultural chemical yields in large complex watersheds that have varying soils and land use and management practices over a long period of time (Borah et al., 2006).
- WARMF: Watershed Analysis Risk Management Framework is currently implemented in a windows-based interface developed by Systech Engineering, Incorporated.

5.3 Model Calibration and Uncertainty

5.3.1 Calibration of Models

Model calibration involves the adjustment of various parts of the model, including input parameter values, within plausible ranges so that the model is a good representation of the real system (Hill, 1998). If model simulations closely match field measurements, then the model is said to be calibrated. The difference between simulated and measured values is often referred to as a residual (e.g., difference between simulated head and measured head in a groundwater flow model). The goal of calibration usually is to minimize a calibration criterion, such as:

- mean error (ME): mean difference between measured and simulated parameters
- mean absolute error (MAE): mean of the absolute value of the differences in measured and simulated parameters
- root mean squared (RMS) error: average of the squared differences in measured and simulated parameters (Anderson and Woessner, 1992)

Calibration may be achieved by trial-and-error or by using an automated parameter estimation program. The universal inverse code (UCODE) is an example of an automated parameter estimation tool that minimizes the residuals using non-linear regression (Hill, 1998). The user guide for UCODE can be obtained using the information provided in Appendix C.

5.3.2 Including Uncertainty in Model Output

Uncertainty is defined as the estimated amount by which a calculated value may depart from the true value. Uncertainty in model output can be due to 1) estimation errors in model input parameters; 2) model algorithms; 3) spatial heterogeneity; and 4) initial and boundary conditions (Shirmohammadi et al., 2006). Most often, uncertainty in model output is assumed to be the result of estimation errors in model input parameters; thus, most reports of uncertainty focus on only this element. The amount of uncertainty in model output is best understood after evaluating the sensitivity of the model output to changes in input parameters. This is accomplished by completing a sensitivity analysis, which is described as systematically changing the input parameters within a plausible range while documenting the average measure of error selected as the calibration criterion (Anderson and Woessner, 1992). In other words, by measuring the effect that uncertain input parameters have on model output.

Uncertainty is usually expressed as a variance from mean, probability (of failure), confidence intervals, or other descriptive statistics (Shirmohammadi et al., 2006). Uncertainty can be reported by automated calibration tools such as UCODE. In the case of UCODE, these uncertainties are reported as:

- Confidence intervals: uncertainty in the simulated values due to propagation of uncertainty in the estimated input parameters
- Prediction intervals: uncertainty expressed as confidence intervals, but includes the effects of measurement error for measured calibration target values (Poeter and Hill, 1998).

Measures of uncertainty for a model are valuable for two reasons. First, model uncertainties can be used to dictate future data collection efforts (e.g., collect more data for parameters with the greatest uncertainty). Second, decision makers can associate an inherent risk to the model results based on the amount of uncertainty.

CHAPTER 6.0

MODEL SELECTION

6.1 Basics of Model Selection

Watershed-scale models can be useful tools for assessing the impacts of onsite wastewater systems (OWS), as well as the relative impacts of OWS compared to impacts of other sources of water pollutants, including centralized wastewater treatment plant discharges. In fact, because the hydrology and chemical transport processes at the watershed scale are so complex, generally numerical models are necessary to provide a rigorous assessment of OWS impacts. Of course, simple models are useful for screening tools, or for gaining a basic understanding of how changes in watershed inputs influence watershed-scale water quality.

Selection of the appropriate model depends on the intended use of the model and the desired output. The ideal model for quantitative watershed-scale assessments involving OWS would have the following attributes.

- **Able to simulate both water quality and water quantity issues**. Foremost, OWS issues usually are related to water quality rather than quantity, because the total volume of OWS inputs and withdrawals are small compared to other watershed-scale inputs and discharges. However, quantity considerations cannot be ignored because they may be important in some areas, particularly in arid regions where OWS may be a significant component of the water budget. A good example of an important quantity problem related to wastewater disposal options is groundwater depletion that occurs from pumping onsite drinking water wells where offsite wastewater treatment is used, because the treated wastewater is discharged to a surface waterbody rather than the local groundwater system.
- **Able to simulate pollutants relevant to watershed management**. The most common pollutants associated with OWS water quality impacts are nitrogen (N) (particularly nitrates) and phosphorus (P). Nitrogen is a regulated contaminant in both surface and groundwaters. Phosphorus is not considered toxic, but can cause eutrophication in surface waters. Other important pollutants that have received much attention recently include pharmaceuticals and other trace organic chemicals (e.g., personal care products, etc) and microbial pollutants (particularly viruses). Of course, other pollutants may be of concern, including organic contaminants (e.g., pesticides, solvents, etc), salts (e.g., road salt), metals, oxygen consuming pollutants, and total dissolved solids.
- **Capable of simulating all relevant compartments of the hydrologic system**. OWS pollutants are typically thought of as a threat to groundwater because the OWS discharges to the subsurface. However, groundwater typically discharges to local streams, and therefore surface water is also potentially impacted by OWS, particularly during low streamflow conditions when the stream is primarily comprised of groundwater. High stream flows during precipitation events tend to dilute the impact of OWS pollutants, although pollutants from other sources may increase during storms (e.g., phosphorus bound to surface sediments that is washed into streams).

Ideally, a model used to simulate OWS pollutants in a watershed-scale system would be able to handle all the hydrologic and transport processes discussed above. Of course, typically no single model is capable of simulating all relevant processes. Nonetheless, it is useful to describe important processes that would be included in the ideal model. First, any watershed scale model must be able to accurately simulate the hydrology of the watershed. This includes precipitation, snowmelt, evapotranspiration, infiltration and runoff, stream flow, groundwater flow, and subsurface transport, including advection, dispersion, and possibly reactions. Given that the vast majority of OWS include soil treatment of effluent, to represent OWS accurately, a model must allow for a subsurface dispersal of liquid effluent with a specified chemical composition. It is also important that the groundwater flow component be rigorous for reasons described above. It is useful (but rarely possible) for the model to simulate infiltration and storage of water and chemical components in the vadose zone, although this adds considerable complexity to the model and requires more input data and time for model development. Nearly all publicly available watershed models treat the movement of water and solute through the vadose zone in a simple manner.

- **Include the relevant chemical reactions in soil, groundwater and surface water.** For nitrogen, this includes the transformation of nitrogen once it enters the soil (nitrification from ammonium to nitrate, and denitrification of nitrate to gaseous nitrogen). Sorption of ammonium ion to soils also may be important. For phosphorus, the most important reactions are sorption to soils (i.e., to metal oxides) and precipitation to a typically immobile solid phase. Both pollutants may also be taken up by plants if the rooting depth is sufficient. Organic contaminants, including pharmaceuticals, will undergo biochemical degradation, sorption to soils, and perhaps other loss mechanisms. Metals will undergo sorption and precipitation to a solid phase. Viruses may sorb to soil or air-water interfaces, deactivate (or die), be killed through predation by other microbes, and be filtered by the soil.

The common reactions for all pollutants are sorption, decay or degradation to another chemical species, and irreversible removal from the aqueous phase (e.g., chemical precipitation for P, straining for virus, etc.). These reactions can be very complex, and can include non-equilibrium reactions, non-linear reactions, irreversible reactions, other-than-first-order reactions (e.g., denitrification is sometimes reported as a half-order reaction). However, practically speaking, models designed for application at the watershed scale cannot simulate this level of chemical complexity, nor is it feasible to obtain all the model input parameters required to simulate such complex processes at the watershed scale. It is therefore minimally sufficient to consider the following: linear first-order reversible sorption; and first-order degradation including treating chemical precipitation, irreversible sorption, or other losses as a first-order loss term. One possible exception to this is the chemical precipitation of phosphorus. This reaction is often significantly rate limited. Because precipitation may often be the principal mechanism of P removal in soils, it may be necessary to include this level of complexity in a watershed model for some cases.

In general, for viruses and all inorganic contaminants, a rigorous geochemical model would be ideal, but the data required to obtain input parameters at the watershed scale are typically not available.

- **Able to account for additions and losses of contaminants via the most important mechanisms**. It is necessary to simulate these mechanisms in such a way that model-input parameters could be obtained with reasonable effort and afforded with a typical budget. In addition, because it is usually important to assess the impacts of OWS compared to other pollutant sources, a model recommended for general use should also be able to account for the hydrologic and chemical processes of the most likely non-OWS sources. At a minimum, these include: processes related to agricultural activity; sediment and chemical runoff due to erosion from urbanization or mining; and point source loading to streams (from industrial sources; or upstream wastewater treatment plant loading). Atmospheric deposition (of N, for example) might be an important process to consider in some settings.

Finally, model selection will always be linked closely to the question at hand including the decision(s) being made and the importance of making correct decisions. For example, nearly all models that are good at handling surface-water processes have simplistic representations of groundwater processes. The converse is true for models that excel at simulating groundwater processes.

Even though recent advancements have been made to integrate and simulate the physical interactions between surface and groundwater, very few models can handle contaminant transport and reactions associated with surface- and groundwater interactions. Thus, the models under consideration for assessing the impacts of OWS on groundwater will usually differ from those considered when stream or lake pollution is the primary concern. In addition, the final model selected may differ depending on the contaminant (e.g., viruses versus nitrogen versus pharmaceuticals) because not all models can handle all chemical transport and reaction processes.

Ultimately, a model recommended for general use should be a widely used model that is available in the public domain. These are practical constraints because OWS issues are often politically and/or legally charged. Thus, only those models that have been vetted by significant prior use will be accepted by the citizen, technical and regulatory communities, and will be able to withstand legal or technical challenges. Finally, it is important to emphasize that the type and capability of the model selected should be consistent with the decisions that the modeling is intended to support. For example, if a $20M treatment plant expansion is being contemplated over continued use of OWS, then the decision may warrant spending $0.5M on modeling studies to assess the relative watershed-scale impact of the decision on water quality and quantity issues.

Appendices A and B provide reviews and descriptions of mass-balance, GIS, and watershed models that have been widely used for watershed-scale assessments, and our analysis of the appropriateness of each model for general watershed-scale simulation of OWS impacts. Appendix C provides a list of Important Documents Associated with this Guide.

6.2 Factors to Consider when Selecting an Assessment Strategy

If a model is necessary to understand current or anticipated nutrient loads from OWS at the watershed-scale, then the selection of the model type becomes critical to planning and management of the watershed. The type of model should be selected by first clearly defining the nutrient and OWS-related question or concern, then selecting the most appropriate type of model. The selected model type can be used to cost-effectively answer the nutrient and OWS-related question within an acceptable level of risk. Models and modeling approaches for evaluating the effects of OWS on water quality can be classified into six primary types of models:

A. Mass-balance screening models are normally implemented in a spreadsheet and are highly simplified estimates of water and nutrient balances. Example models include P&NLRM and OSF.

B. GIS-based screening models are implemented either completely within a GIS or require the use of GIS for processing of geographically-referenced data (e.g., land-use, soil property and precipitation) as primary model inputs. Example models include NLM, MANAGE, and PLSM.

C. Surface-water models focus on fluxes between the atmosphere and ground-surface and changes in storage within rivers, streams and other surface waterbodies. Normally a combination of empirical models and/or numerical models that can predict flow rates and nutrient concentrations over various time-scales. Example models include HEC-HMS, HSPF.

D. Vadose zone models are built around the unsaturated zone and consider fluxes between the overlying ground surface and the underlying saturated zone. An example model is HYDRUS.

E. Groundwater models focus on changes in storage and fluxes within the saturated zone. Normally a combination of empirical models and/or numerical models that can predict hydraulic heads and nutrient concentrations over various time-scales. Example models include MODFLOW, FEFLOW, SUTRA.

F. Integrated watershed models include surface water, vadose zone, and groundwater flow modules; allowing for temporal calculations of flux and storage in each component of the hydrologic system. Example models include GWLF, MIKESHE, SWAT, WARMF.

The types of models listed above are arranged, generally, in order from least complex (A) to most complex (F). Increased complexity normally translates to higher costs for gathering of necessary data, implementation and calibration of the model. So, in all cases, it is suggested that the watershed-scale OWS question be analyzed with the simplest model type (Model type A) and progress to more complex model types as necessary. Model type A usually produces a conservative estimate of nutrient loads, which allows the user to either be: 1) satisfied that there is not a problem (due to conservative estimate) and end the modeling process; or 2) unsure of severity of problem so continue refining the "answer" by implementing more complex model types. The types of models suggested below would be the most complex required considering various factors and conditions of the study watershed.

6.3 Factors to Consider in Model Selection

When selecting the appropriate type of model for watershed assessment, the primary factors to evaluate are: 1) status of watershed; 2) regulatory and environmental issues; 3) hydrologic, chemical, and physical processes; 4) relative costs of a model; and 5) information needs.

6.3.1 Status of Watershed

Does the watershed have an existing water quality problem related to OWS or is the modeling process required to plan for land-use change and a significant change in the number of OWS in the watershed?

An evaluation of the status and trends in nutrient concentrations should be completed to understand the relationship between current water quality and established criteria. Increasing concentrations (worsening) of nutrients or other indicators would suggest a need for a more complex model. In these cases, a more rigorous model could model water quality conditions under current land use and wastewater disposal practices, while predicting the outcomes of alternative practices. Using available data, the status and trends in water quality should be qualitatively evaluated for both surface and groundwater to determine if the status of the watershed is:

- unknown impairment or insufficient data (Model type A);
- no impairment, or few if any values above environmental criteria (Model type B);
- little impairment, and expected increase in nutrient levels due to additional OWS (Model type C, D, E, or F);
- moderate impairment, or small percentage of values > criteria (Model type C, D, E, or F); or
- severe impairment, or large percentage of values > criteria (Model type C, D, E, or F).

6.3.2 Regulatory and Environmental Issues

What water quality problems (resulting from OWS) cause public concern? Are water resources impaired relative to regulatory standards?

The consequences of impairment from OWS also influence the selection of an assessment strategy. A few of the most common consequences include:

- Economic loss (Model type A)
- Aesthetic value (Model type B)
- Degraded aquatic habitat (Model type C, D, E or F)
- Human health hazard (Model type C, D, E, or F)
- Regulatory action (Model type C, D, E, or F)

If the surface or groundwater at risk from OWS contamination is also a source of drinking water in the watershed, there may a significant human-health risk. Resources and thus consequences are not independent and if one resource is impacted it may lead to several consequences. For example, OWS contamination of a shallow aquifer used for drinking water might result in regulatory action requiring advanced treatment OWS or reduced development density, either of which would cause economic loss. If the shallow aquifer discharges nutrients to streams via base flow, aquatic habitat might be degraded, aesthetic and recreational values might be lost, and economic losses from tourism might occur.

In the special case where there is already active management or regulatory restrictions in effect, such as TMDLs, 303D listings, or water management area designation, it can be assumed that water resources are severely impaired and a more complex modeling approach may be warranted.

6.3.3 Water Resources at Risk

Which part of the hydrologic system is most affected by OWS effluent? Which is the receptor of nutrient loading? Are the surface water and groundwater systems well connected and equally at risk? What processes dominate fate and transport of nutrients in the watershed?

In areas served by OWS, groundwater is typically an important source of drinking water. Where homeowners supply their own water from individual wells, the source is usually a shallow aquifer that would be vulnerable to contamination from OWS. If there is a community water system, wells may tap deeper aquifers that are less likely to be impacted by OWS effluent. In the latter case, it might not be necessary to use a model that explicitly simulates groundwater flow and nutrient transport. However, if the source of drinking water is a shallow (water table) aquifer, evaluation of alternatives for protecting the aquifer would likely require a model capable of simulating the distribution of nutrient concentrations in the subsurface.

Where surface water is a primary source of drinking water, or where aquatic habitat is a resource at risk from OWS contamination, models capable of simulating surface water and soil processes might be most appropriate.

Examples of receptors of nutrient loading and dominant processes in the hydrologic system include:

- Stream, lake, or reservoir used for non-potable municipal, industrial, agricultural, recreational, and other purposes (Model type A or B)
- Stream, lake, or reservoir providing drinking water (Model type C or F)
- Groundwater used for non-potable municipal, industrial, agricultural, recreational, and other purposes (Model type A or B)
- Groundwater providing drinking water (Model type E or F)
- Surface water and groundwater providing drinking water (Model type F)

6.3.4 Relative Costs of a Model

Are there budgetary limitations for assessing impact of OWS on the watershed? Do the budgetary limitations allow for an adequate level of risk assessment, or should the budget be supplemented to allow for more complex and risk-based modeling?

Financial resources may limit the choice of model type to a more simplistic and inexpensive model. If human or animal health is at risk, however, a complex (and thus generally expensive) model is necessary regardless of cost. In other words, if the resource at risk is groundwater or surface water providing drinking water, possible consequences include human health hazard. In such cases, the damage costs cannot even be estimated in monetary terms. Users or regulators should implement more complex models to evaluate possible abatement scenarios, which would be still far cheaper than the damage costs.

Under other conditions, simpler models could provide a conservative estimate. Relative cost of the model is directly related to time involved in gathering of necessary data, implementation and calibration of the model.

Relative costs range from negligible to high:

- Negligible cost (Model type A)
- Low cost (Model type B)
- Moderate cost (Model type C, D, or E)
- High cost (Model type F)

Modeling costs are determined by considerations such as: how long it takes to set up, calibrate and apply the model including data gathering (and whether the data are available at no cost from public source); how much training is needed; how much the model itself will cost; and whether support is available during model setup and calibration. Some models require substantial training or modeling expertise, especially if the model is a complicated model. Some may require moderate training and others may require very limited training if the modeler has familiarity with basic environmental models. The required training is reflected in the overall cost of a model. Data requirements also affect modeling costs. Actual costs to purchase a modeling program may vary. Some models are freely available while others may cost several hundred or thousands of dollars.

A subjective assessment of watershed models is shown in Table 6-1. Factors include costs for training a model user; model operation (model setup and calibration); cost of data collection; level of model complexity; and corresponding level of information generated. Actual costs could vary depending on who does each of the modeling tasks. In some cases model developers can be contracted to setup, calibrate, and complete scenario evaluations. In other cases, a client can complete the calibration and scenario evaluation after the model is set up by the developer. Models could be relatively inexpensive and less complex generating a limited amount of information or could be relatively expensive and more complex generating more information. The best alternative model would be a simple model that costs less in terms of all cost components (set up, calibration, training, etc.) and at the same time is able to generate sufficient and accurate information.

Table 6-1. Modeling Cost, Complexity, and Level of Information Generated.

Model Type	Cost and complexity					
	Training Cost	Operation Cost	Cost of Data	Purchasing Cost	Level of Complexity	Information Generated
SWMM	L	L	L	Free/Public	L	L
OSF	L	L	L	Free/Public	L	L
PLS	L	L	L	Free/Public	L	L
NLM	L	L	L	Free/Public	L	L
MANAGE	L-M	L-M	L-M	Free/Public	L-M	L-M
GWLF	L-M	L-M	L-M	Free/Public	L-M	L-M
GLEAMS	M	M	M	Free/Public	M	M
ANSWERS	M	M	M	Free/Public	M	M
CREAMS	M	M	M	Free/Public	M	M
AGNPS	M-H	M-H	M-H	Free/Public	M-H	M-H
WARMF	H	H	H	Free/Public	H	H
SWAT	H	H	H	Free/Public	H	H
HSPF	H	H	H	Free/Public	H	H
MIKE-SHE	H	H	H	Proprietary	H	H

Low=L, Medium=M, High=H

6.3.5 Information Needs

What data are available for the watershed (e.g., population data, OWS data, nutrient measurements, groundwater levels, hydraulic parameters, surface water discharge)? Are these data available over long periods of time? Are they geographically distributed?

In some watersheds, data has not been collected systematically and lacks geographic coverage, periodic measurements over time, and consistency in how samples were collected, analyzed and reported. When other criteria indicate that a more complex model should be implemented, users may opt for a simple model formulation with fewer data requirements and lower time/cost for implementation. The data available for assessing the watershed may determine which model type is possible according to the following guidelines:

- Very limited data – population, minimal water quality and discharge measurements are available (Model type A)
- Limited data – spatially distributed land-use, topography, and soils data are available (Model type B)
- Moderate data – detailed data are available for the surface water system such as discharge measurements and water quality over time (Model type C)
- Moderate data – detailed data are available for the unsaturated zone such as soil and unconsolidated material physical properties (Model type D)
- Moderate data – detailed data are available for the groundwater system such as water levels, hydraulic properties and water quality (Model type E)
- Abundant data – detailed data are available for the surface water and groundwater systems such as water levels, hydraulic properties, discharge measurements and water quality over time (Model type F)

6.3.6 Pollutant Fate and Transport Processes

An additional factor that must be considered is whether the model selected can sufficiently account for the pollutant fate and transport processes needed to address the problem at hand. For example, if the model selection process presented in Section 6.4 suggests that an integrated watershed model should be used, care must be taken to select a model that can handle the pollutant(s) of interest. For example, some models are well suited to handle nitrogen transport, but are not well suited to handle phosphorus transport. Incorporation of pollutant fate and transport processes in model selection is described in more detail in Section 6.5.

6.4 Using the Model (Type) Selection Matrix

When selecting the appropriate model type, it is suggested that the watershed-scale OWS question be analyzed with the simplest model type (Model type A) and progress to more complex model types as necessary (top of Figure 6-1). Model type A usually produces a conservative estimate of nutrient loads, which allows the user to either be: 1) satisfied that there is not a problem (due to conservative estimate) and end the modeling process; or 2) unsure of severity of problem so continue refining the "answer" by implementing more complex model types. Figure 6-1 shows the factors to consider when selecting a model type, and indicates the most complex model that is appropriate under conditions of the watershed. This model type selection matrix (Figure 6-1) can be used to understand the complexity of modeling required to answer the questions about watershed management and OWS.

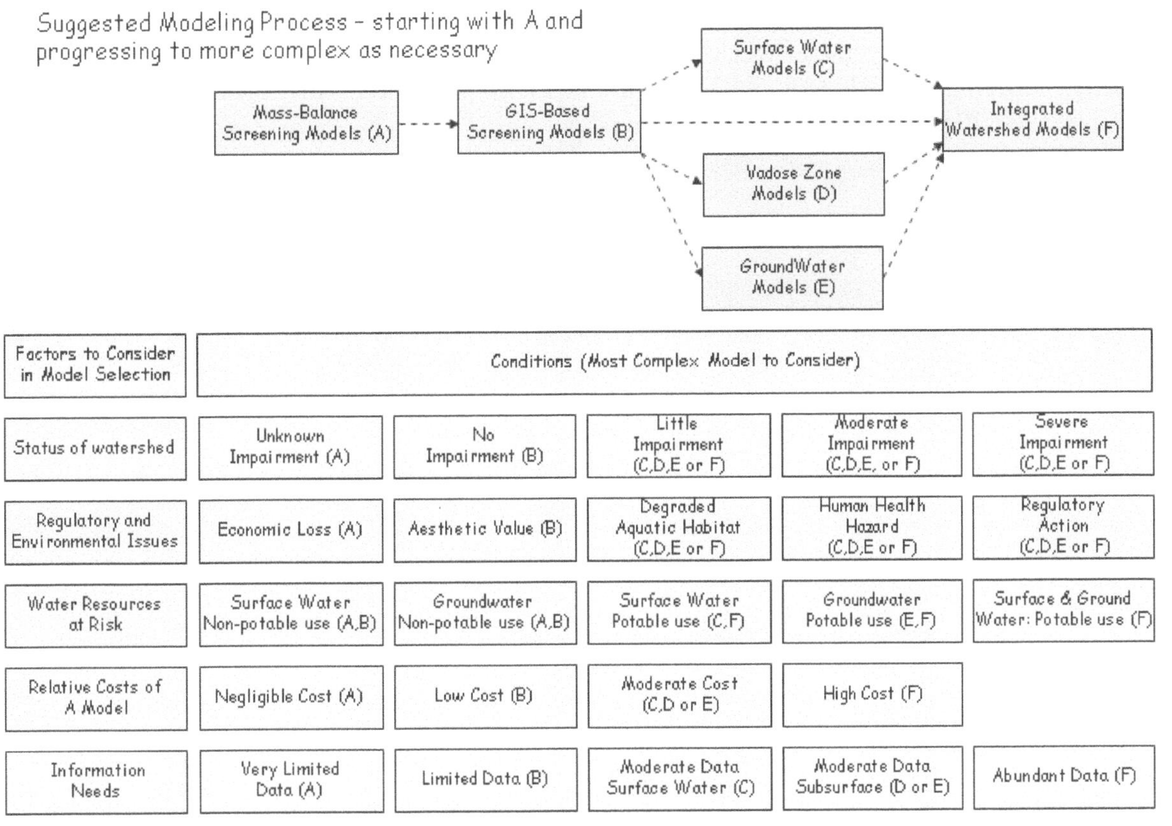

Figure 6-1. Model Type Selection Matrix.

6.5 Selecting Models that Incorporate the Appropriate Pollutant Transport Mechanisms

Once the appropriate type of model has been chosen, the next step is to choose a particular model that can handle the transport and transformation of the pollutants that are of concern (recall Chapter 3.0). For example, some models do a very good job simulating nitrogen transport, but do not have rigorous formulations for phosphorus transport. Some models that can simulate N and P transport and transformations cannot handle microbial transport. In the researchers' experience, it is usually one of these pollutants that drive the decision-making process regarding OWS. If more than one of these pollutants are considered to be important with regard to the decision making process, then it is likely that more than one model must be implemented. Unfortunately, in some cases, it will not be possible to collect the data required to implement a particular type of model. Then, models that have simplified formulations for transport of one or more pollutants may be used, but the results of that model cannot be considered as reliable as one that more accurately represents the true fate and transport processes.

At the time of publication of this report, very few watershed models can handle a variety of pollutant transport processes as well as processes relevant to OWS. In particular, the ability to simulate discharge of aqueous septic tank effluent containing various wastewater constituents is rare among models. The researchers currently recommend the WARMF model (see Section 5.2) for pollutant transport modeling at the watershed scale. However, the SWAT model (also described in Section 5.2) is currently being revised to enable better incorporation of OWS (Dr. C. Santhi, Texas A&M University, Blackland Research Center, Temple, Texas 2008). For problems primarily involving groundwater contamination at a sub-watershed scale, the researchers recommend Modflow-MT3D/RT3D (Section 5.2).

CHAPTER 7.0

OBTAINING DATA FOR WATERSHED MODELS

7.1 Introduction

A watershed model represents a set of equations and numerical and logical steps that converts numerical inputs to numerical outputs. Watershed models require a diversity of information in order to run. The information includes both model input variables and parameters. (The latter refer to numerical constants in the equations that dictate the quantitative behavior.) Model input variables are independent and measurable quantities with a definite numerical value (e.g., precipitation, temperature) while parameters are physical quantities with a numerical value or a value within a certain range, which can be measured, estimated or calculated via calibration and that determine how model input variables will be transformed into output.

All models require some basic inputs, including climatic variables, and nutrient sources. Most models that assess OWS impacts must also account for sources of potential pollutants from other sources (e.g., atmospheric, overland flow to streams, infiltration to groundwater, from other land uses, etc.). There is a strong relationship between model complexity and input-data requirements. Model types vary from simple screening models with much less data required to more complex physically based models that require extensive data. Much of the model-input data required can be found from governmental sources, and most of these data are available on the internet, although "hard-to-get" paper sources can be invaluable.

This chapter presents basic input data required to set up a model, and suggests possible databases or sources for these data.

Chapter 8.0 provides information on model parameters that can be measured, estimated or determined via calibration required to model hydrology, sediment, and the most relevant OWS constituents (e.g., nitrogen, phosphorus and organics). Parameters control the behavior of transformation of the inputs into outputs such that the outputs reflect observed natural conditions. Once the model is set up, model parameters are adjusted to reflect the watershed characteristics and to yield a better match between model outputs and observed data. Figure 7-1 shows a general flow chart to model hydrology or any other constituent including required inputs.

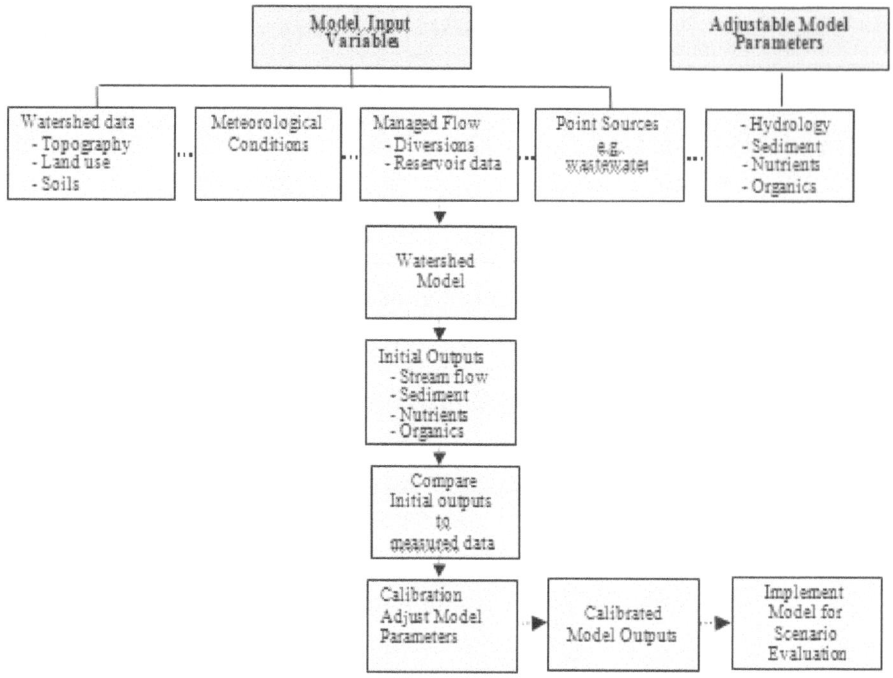

Figure 7-1. WARMF Input Variables, Parameters, and Modeling Process.

7.2 Data Inputs for Different Model Types

7.2.1 Mass Balance Screening Models

Mass balance screening models are simpler models that use data available in tabular form and may not contain specific geographical locations. Examples include: 1) the Population and Nutrient Level Regression Model (P&NLRM), which is a simple regression model for nutrient level (load/person/time) at the watershed outlet calculated based on flow rate, concentration, and population; and 2) the On-Site and Fertilizer (OSF) model developed by Gaines (1986) that can be used to calculate nitrogen load based on population, septic load, and fertilizer load per household per unit time. These models use average annual values for discharge and loads because data are most often available on this time interval. Each data input variable should be representative of the entire watershed.

7.2.2 GIS-Based Screening Models

GIS-based screening models can be built from a combination of tabular information and geographically referenced data (i.e., GIS data) or entirely within a GIS from datasets that are geographically referenced. These models require GIS data such as digital elevation models (DEM), land use-land cover data and soils data.

7.2.3 Surface-Water Models

Surface water hydraulic and water quality models include the Enhanced Stream Water Quality Model (QUAL2E), hydrodynamic and water quality model (CE-QUAL-W2), the Environmental Fluid Dynamics Code (EFDC), Hydrodynamic-Eutrophication Model-3D (HEM3D) models, Water Quality Analysis Simulation Program (WASP) and several others. These models can be one dimensional (QUAL2E), two-dimensional (CE-QUAL-W2), or three-dimensional (EFDC/ HEM3D). Some can be applied in one, two, or three dimensions (e.g., WASP).

Most of these models require definition of the domain, inflow, outflow and water quality boundary conditions, and meteorology data. The domain data that may include configuration of branched stream network, cross-sectional profiles or geometric representations of cross-sectional area, located as required to define changes in channel geometry, and tributary angles at junctions. Bathymetric data can be obtained from regional water districts or can also be generated using USGS quadrangle maps. Bathymetry data can be used to derive the average segment width, depth, and orientation information, along with bottom roughness and initial water surface elevation for river segments. The flow data that may be required are upstream discharge or elevation, downstream discharge, elevation or rating curve, time-dependent lateral inflows, time-dependent point inflows or withdrawals.

The meteorology data may include cloud cover, wind speed, dry bulb temperature, wet bulb temperature, and atmospheric pressure. The water quality inputs could be upstream boundary values for all constituents modeled and lateral inflow values at each segment. Meteorological data are an important component of surface-water models since they determine the surface boundary conditions. The meteorological data such as air temperature, dew point temperature, wind speed, wind direction, and cloud cover can be obtained from the National Climatic Data Center (NCDC).

They also may require specification of inflows for branches, tributary inflow (e.g., point sources), and distributed inflow (e.g., nonpoint sources). These models also require specifying initial conditions for the water quality state variables. The initial concentration of each active state variable is estimated based on the general magnitude of the in-stream monitoring data.

The output from other watershed models can be used as input to the hydrodynamic models. These outputs from other watershed models may include runoff, outputs sediments and water quality variables, such as total suspended solids (TSS), nitrogen, and phosphorus species. The simulated daily average outputs from the watershed models can be used as boundary conditions that are input into surface-water models.

7.2.4 Vadose-Zone Models

These models compute water flow and solute transport in variably saturated porous media. They can be simple or more complex finite element model such as HYDRUS 2D/3D. The inputs required for flow are initial conditions, boundary conditions and soil properties. For solute transport, inputs on solute initial concentration and boundary conditions are required in addition to data required to run flow simulations. The flow and solute boundary conditions are more important than the initial conditions. Flow and solute boundary conditions are presented in Section 7.3.

7.2.5 Groundwater Models

These are models that can be used for fate and transport of contaminants in the saturated zone. An example is MODFLOW, the general three-dimensional groundwater model developed by the USGS (Hill, 1992). It can be used in conjunction with other models (e.g., MT3D) to simulate changes in contaminant concentrations such as nitrates or phosphates in groundwater considering advection, dispersion, and chemical reactions. Information required to set up a MODFLOW model includes layer types and cell attributes such as storage coefficients and transmissivity, river type boundary conditions, recharge to the groundwater from precipitation, well type boundary conditions, drain type boundary conditions, evapotranspiration and head type boundary conditions and initial conditions. Some of the data sources include observation wells, pump test/slug test data, and drillers' report. Mass balance models (precipitation minus evapotranspiration) can calculate recharge data. Chemical loading from the vadose zone can also be estimated using a vadose zone model such as HYDRUS or a simple mass balance model.

7.2.6 Integrated Watershed Models

Physically-based watershed models such as the Soil and Water Assessment Tool (SWAT) and Watershed Analysis Risk Management Framework (WARMF) account for all of the known processes for simulating the water cycle and nutrient cycle based on climate, topography, soil properties, land use, and management practices. DEM, land use/land cover data, soils data, and weather data are required to build these models. The DEM is used to calculate parameters such as slope and slope length as well as the definition of stream network.

These models are useful primarily for simulating transport of a contaminant once it enters streams and lakes – including interactions with the bottom sediment, suspended sediment, and the biota in the water – but do not account for any watershed processes outside the stream. For the most part, the watershed hydrological and water quality models, such as SWAT, HSPF and WARMF, have the primary surface water components needed to simulate pollutant transport in streams, although they are not as rigorous as some of the surface water models.

These watershed models are most suited to simulate the processes that take place in the upland watershed and narrower streams where 1-D hydrodynamic and water quality assumptions are valid. These models may not be appropriate to simulate the hydrodynamics and water quality processes of larger waterbodies such as lakes, reservoirs, and bigger streams where 2-D or 3-D computations may be required. The hydraulic and water quality models are suited to tackling the hydrodynamics and water quality processes in larger waterbodies and do not consider upland watershed processes.

7.3 Data Sources for OWS

7.3.1 Population and Flow Rate Data

The most common source of data for geographic location of onsite wastewater systems (OWS) is the GIS data associated with the 1990 census. Population data can be obtained from the U.S. Census Bureau (USCB) via the web-site http://www.census.gov/.

More recent data than 1990 are necessary when evaluating the effects of population growth on nutrient loading, which is often a reason for watershed-scale assessments. The recently released 2000 USCB data does not include a geographic distribution of septic systems at the watershed scale. Thus, model users need other local sources of information on distribution of septic systems. For instance, if GIS data are available from county tax records, it can be used to represent OWS distribution with points located at the centroid of each tax lot. The distribution can then be used to calculate the number of people in each catchment.

The average number of people per septic system/tax lot is about 2.5 according to the USCB. The average STE flow per person is about 60 gallons per capita per day based on earlier studies by the U.S. EPA and a recent review by McCray et al. (2005), but this is an uncertain number. For averaging large numbers of OWS over a watershed unit, such an average is probably appropriate. However, one may wish to be more or less conservative. Hydraulic loading rates in the range of 0.24 cm/d (Robertson and Cherry 1995) and 19 cm/d (Weiskel and Howes 1992) are reported in the literature. The mean and median values reported in Weiskel and Howes (1992) are 3.5 cm/day and 3 cm/day. These values can be used to specify flow boundary conditions.

Figure 7-2 shows a cumulative frequency diagram (CFD) of flows based on the literature review of more than 50 sources by McCray et al. (2005). The 50% frequency is the median rate reported, the 80% frequency value for flow is about 90 gal/cap/day, meaning 80% of values reported in the literature were less than this value. The 25% flow rate is about 42 gal/cap/day, which means that 75% of all reported values were more than 42 gal/cap/day. The 25 and 75 percentile values are also called "quartile" values.

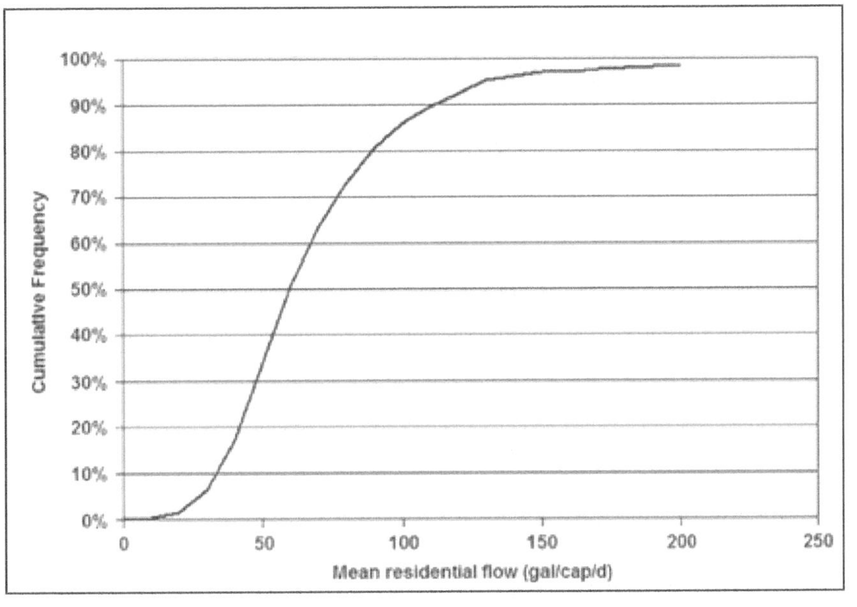

Figure 7-2. Cumulative Frequency Diagram for Septic Tank Flow.
From McCray et al., 2005. Reprinted from *Ground Water* with permission of the National Ground Water Association, ©2005.

7.3.2 Pollutant Transport and Transformation Parameters

Many other parameters are required for pollutant transport and transformations. This information is highly uncertain, and must be obtained either via calibration or via uncertainty analysis. Summary of model input data from McCray et al. (2005) is given in Table 7-1. N and P concentrations in septic tank effluent are given in Table 7-2. Data related to OWS nitrate transformation (first order nitrification and denitrification rates) in soil treatment units is reported in Table 7-3, from McCray et al. (2005). CFD diagrams for N and P concentrations in STE shown in Figure 7-3 and Figure 7-4 can be used to define solute boundary conditions.

Table 7-1. Summary of Model Input Data from McCray et al. (2005).

Phosphorus concentrations in STE	CFD, range, median, n [1]
Phosphorus sorption coefficient	CFD, range, median, n
Maximum P sorption capacities by soil type	Range, median, n
Ammonium-N concentrations in STE	CFD, range, median, n
Nitrate-N concentrations in STE	Range, median, n
First-order denitrification rate	CFD, range, median, n
First-order nitrification rate	CFD, range, median, n
Zero-order nitrification Rate [2]	Range, median, n
Organic N	Range, median, n
Total N	Range, median, n
Wastewater flow rate	CFD, range, median, n

[1] CFD = cumulative frequency diagram, n = number of data

[2] Tucholke et al. (2007) and Tucholke (2007) provide detailed information about zero-order denitrification rates for various soil types and conditions

Table 7-2: Summary of N and P Concentrations in STE.

Pollutant	Concentration (mg/l)			
	Median	Range	Number of Data	Average Value fromU.S. EPA (2002)
TN (mg N/L)	52 or 68	12 to 453	18	44.2
Organic N (mg N/L)	14	9.4 to 15	6	--
Ammonium (mg N/L)	58	17 to 178	37	--
Nitrate (mg N/L)	0.2	0 to 1.94	33	0.04
Phosphate (mg P/L)	9.0	1.2 to 21.8	35	8.6

Figure 7-5 is a CFD of first-order nitrification rates summarized from literature sources. Denitrification is possible both at anaerobic microsites in the vadose zone and in anaerobic groundwater when a source of carbon is present and available (McCray et al., 2005). While many studies have been performed on nitrate removal in the subsurface via denitrification, it is still neither well understood nor well quantified. First-order denitrification rates vary widely in literature sources. The first-order reaction rates for denitrification are presented in a CFD in Figure 7-6. Due to the large variability in literature data, nitrification and denitrification rates could not be segregated by soil type. Nearly all of the N in STE is in the form of ammonium nitrogen, and then gets rapidly converted to nitrate in the subsurface. More detail on the frequency and distribution of pollutant concentrations reported in STE, is described in Chapter 8.0.

Table 7-3. Summary of Nitrification and Denitrification Rates.

Process/Reaction Order	Median	Range	Number of Data
Nitrification-zero order, μ' (mg/L/d)	264	156 to 1464	7
Nitrification-first order, k_1 (1/d)	2.9	0.0768	19
Denitrification-zero order, k_2 (1/d)	0.025	0.004 to 2.27	53

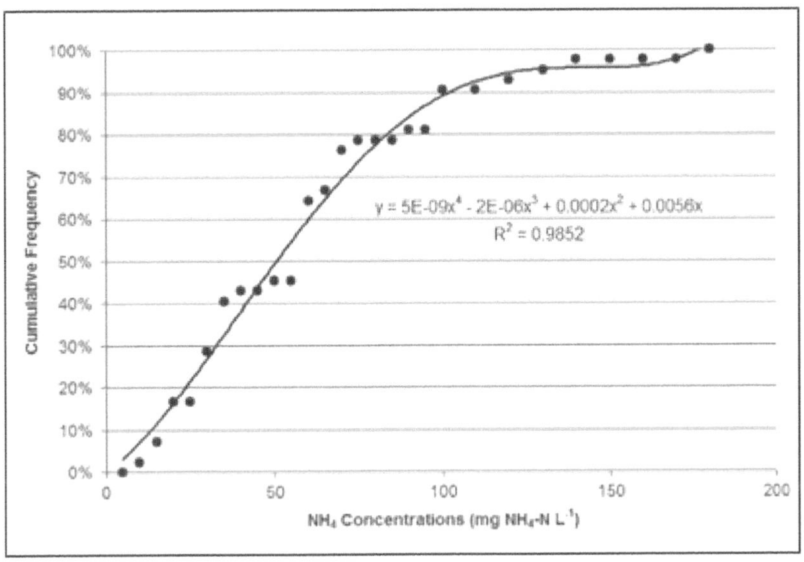

Figure 7-3. Cumulative Frequency Diagram for Ammonium in STE.
From McCray et al., 2005. Reprinted from *Ground Water* with permission of the National Ground Water Association, ©2005.

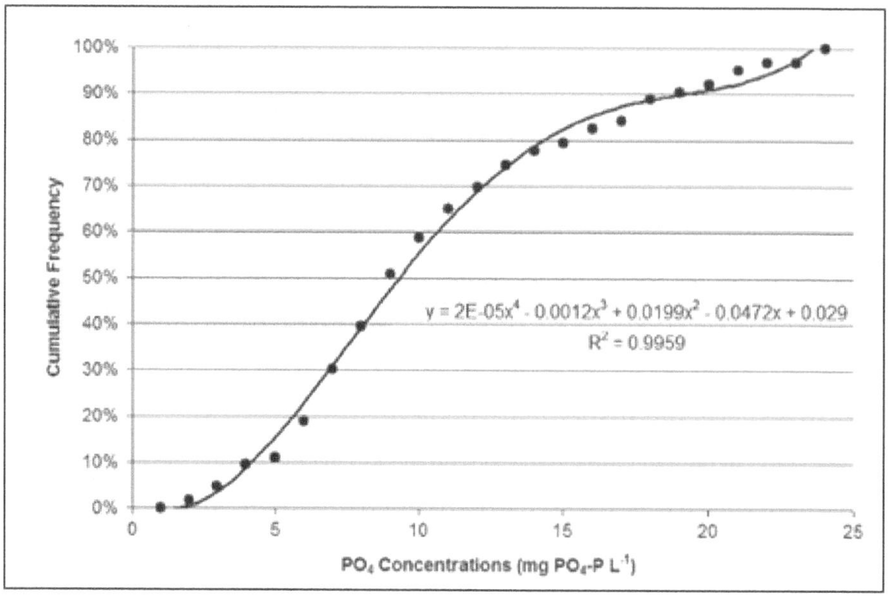

Figure 7-4. Cumulative Frequency Diagram for PO4-P Concentration in STE.
From McCray et al., 2005. Reprinted from *Ground Water* with permission of the National Ground Water Association, ©2005.

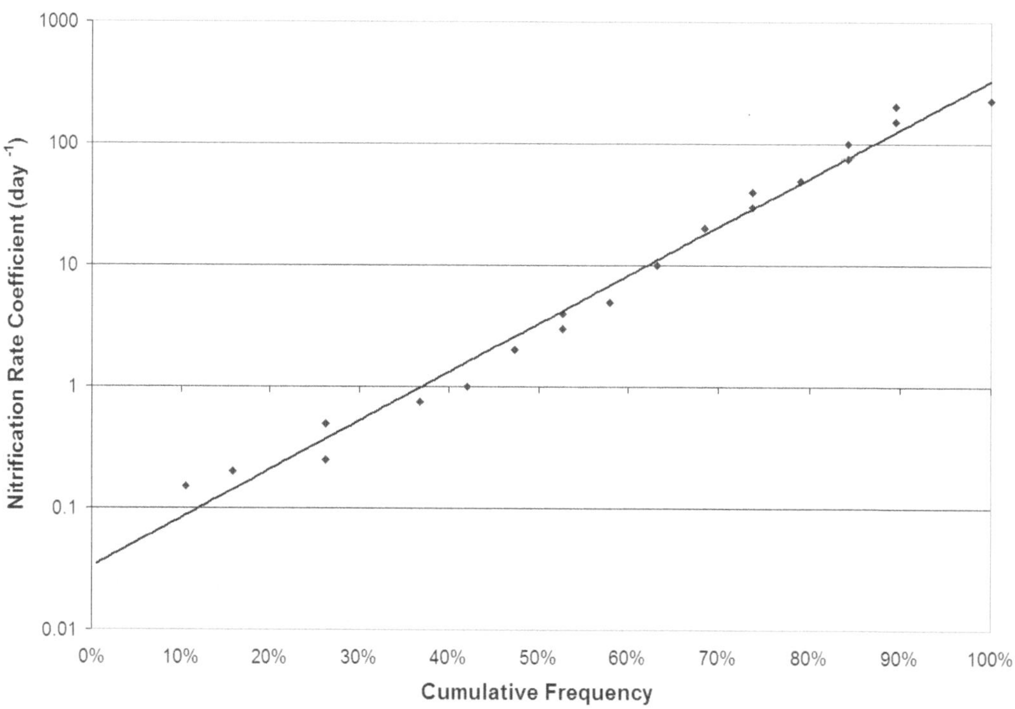

Figure 7-5. Cumulative Frequency Diagram for First-order Nitrification Rate.
From Heatwole and McCray, 2007. Reprinted with permission from Elsevier.
See also McCray et al. (2005) for detailed explanation.

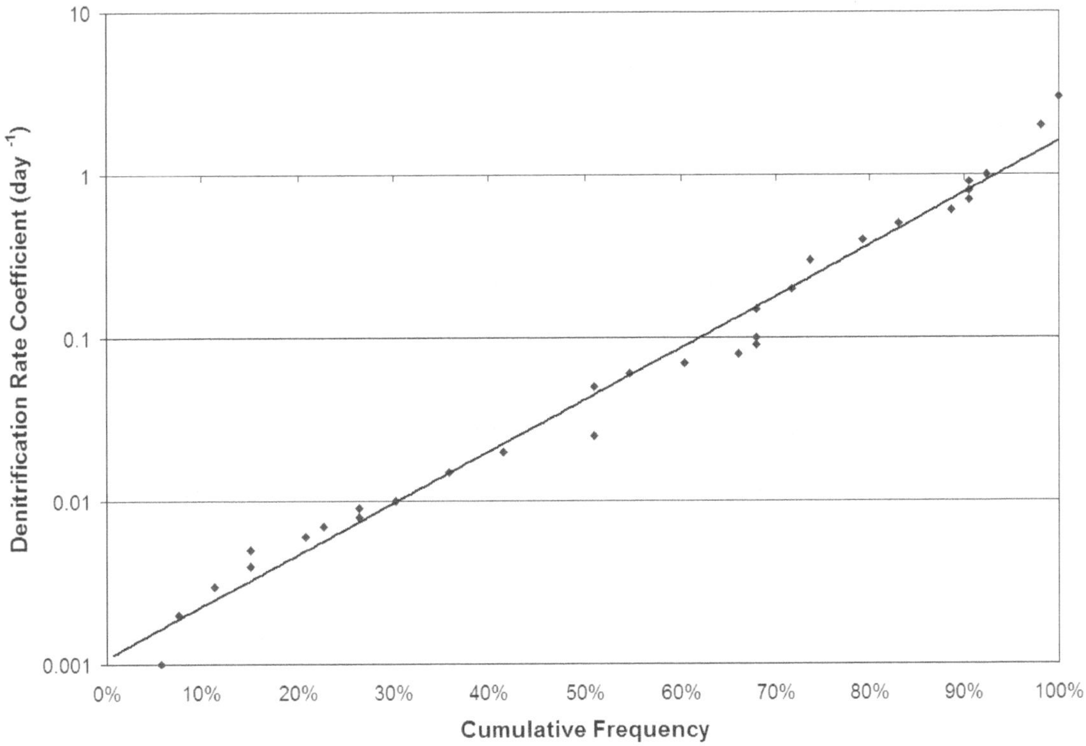

Figure 7-6. Cumulative Frequency Diagram for First-order Denitrification Rate.
From Heatwole and McCray, 2007. Reprinted with permission from Elsevier
See also McCray et al. (2005) for detailed explanation.

7.3.3 Point Sources

Point source data from industries and wastewater plants (flow and load) can be obtained from the EPA Permit Compliance System (PCS). The National Pollutant Discharge Elimination System (NPDES) data files generated using the United States Environmental Protection Agency's (EPA) Permit Compliance System (PCS) database are also available. Monitoring data from NPDES is available at http://www.epa.gov/enviro/index.html.

7.3.4 Sediment

Sediment data are important if OWS impacts are being studied relative to other impacts (e.g., construction, agricultural, recreational land, etc.). These other land uses can be a source of constituents. For instance, phosphorus and organics can bind to sediments and be transported to surface water via overland transport.

7.4 Other Fundamental Model Input Variables

7.4.1 Land Cover (Vegetation and Land Use)

Land cover data are required for most surface water models, as land cover is the most important factor affecting the amount of runoff and nutrients from the landscape. Vegetation has a large influence on the movement of water and nutrients over the land surface, into groundwater, and out of groundwater through evapotranspiration. The vegetative cover of an area of land is typically inferred from GIS data describing the land cover or land-use characteristics.

The Multi-Resolution Land Characteristics (MRLC) consortium is a group of federal agencies, including the USGS, EPA and National Oceanic and Atmospheric Administration (NOAA), which works to maintain a nationally consistent dataset of satellite and remote sensing land use data. A National Land Cover Dataset (NLCD) of the entire conterminous United States was developed by the Consortium for the year 1992. These data are also available online via http://seamless.usgs.gov/. A current effort, beginning in 1999, is underway to develop a NLCD representing land use for the year 2001.

Another common Land Cover dataset is the Land Use/ Land Cover (LULC) dataset which was developed by the USGS in the early 1990s and is based primarily on aerial photography from the 1970s and 1980s.

The LULC data, commonly referred to as GIRAS land-use data, are distributed mostly in the form of shapefiles (vector GIS data) while the NLCD data are typically distributed in a raster form. The BASINS model can be used to automatically download a theme containing the EPA GIRAS LULC land use. These data are relatively coarse and represent prior land use and land cover but can be useful for most projects. Figure 7-7 below shows a land-use map constructed for a watershed model.

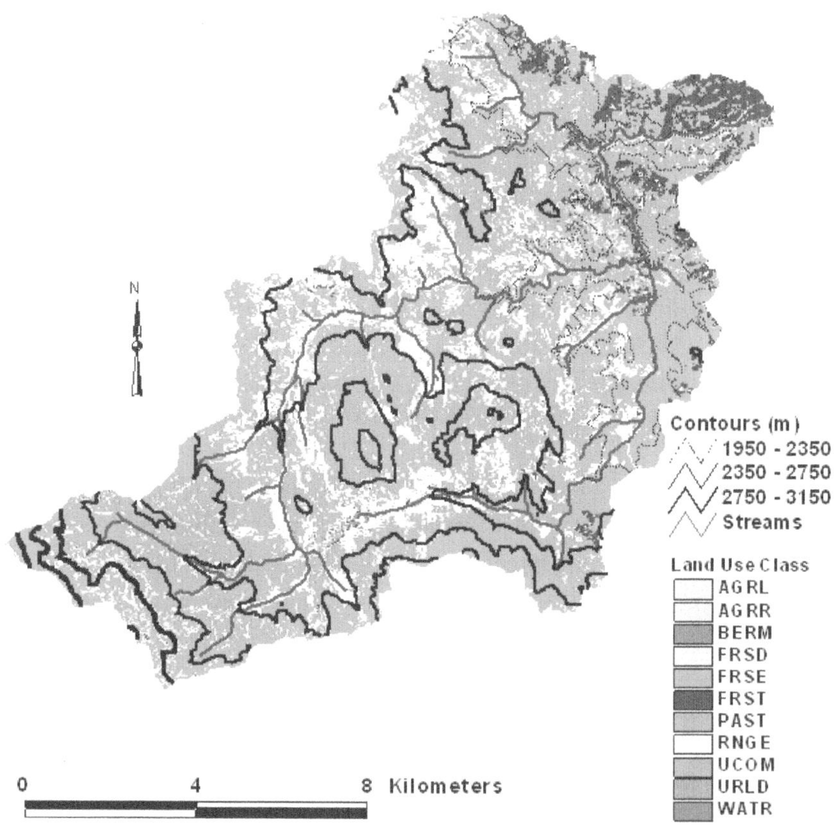

Figure 7-7. Land Use Map for Turkey Creek Watershed, Colorado.
Geza and McCray, 2008. Reprinted with permission from Elsevier.

7.4.2 Soil

Soil type is another characteristic commonly used when modeling watershed scale processes. The Natural Resource Conservation Service (NRCS) is responsible for collecting and distributing soil survey information for the United States via the U.S. Department of Agriculture Data Gateway http://datagateway.nrcs.usda.gov/ (see Table 7-3). The two NRCS datasets most widely used for natural resource planning and management are the Soil Survey Geographic Database (SSURGO) and the State Soil Geographic Database (STATSGO). The SSURGO database is the most detailed database available and is intended for use on a county/township/landowner scale. The STATSGO database is a generalization of the SSURGO database and is intended for regional/multistate/river basin scale planning and management. Soils data are necessary for most surface water models to identify soil types and composition in the entire watershed.

Figure 7-8 shows a SSURGO soils map constructed for a watershed model, while Figure 7-9 shows data from the less detailed STATSGO database.

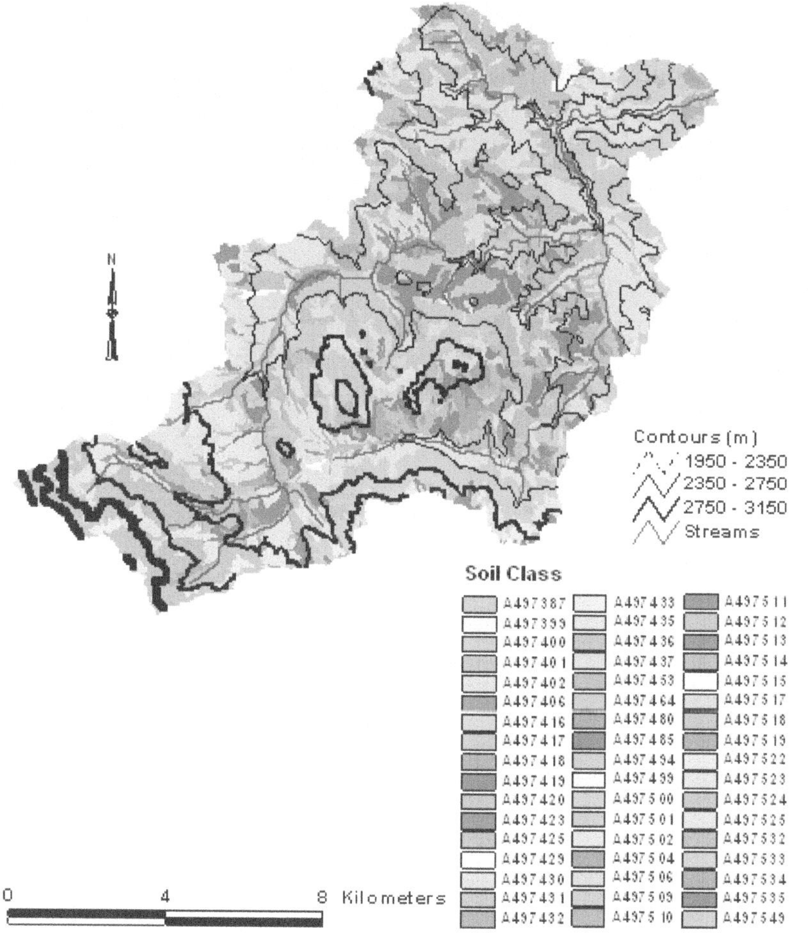

Figure 7-8. SSURGO Soils Map for Turkey Creek Watershed, Colorado.
Geza and McCray, 2008. Reprinted with permission from Elsevier.

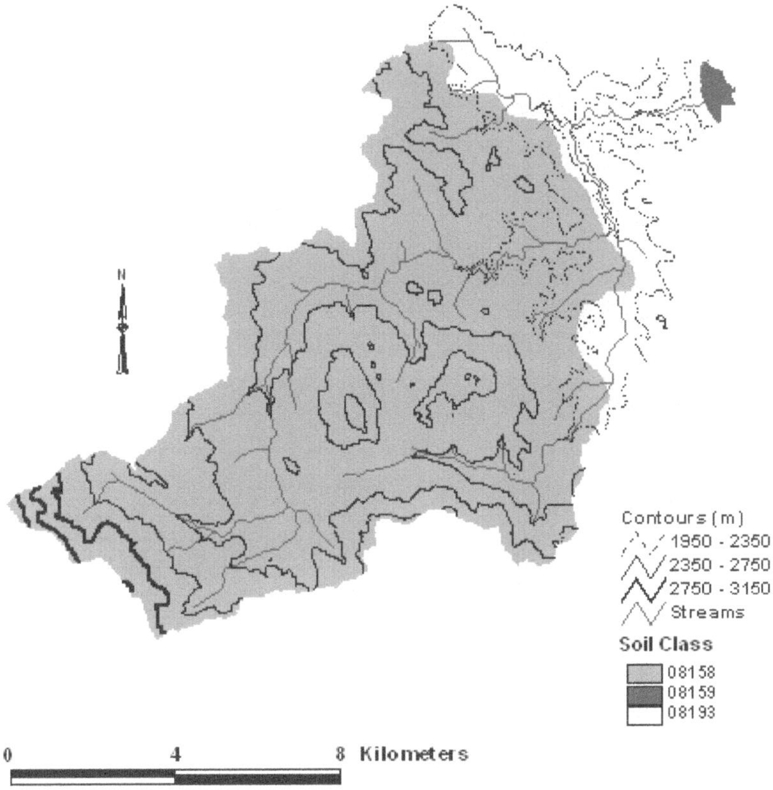

Figure 7-9. STATSGO Soils Map for Turkey Creek Watershed, Colorado.
Geza and McCray, 2008. Reprinted with permission from Elsevier.

7.4.3 Topography

DEM and land cover data can be obtained from the U.S. Geological Survey (USGS) seamless data distribution page http://seamless.usgs.gov/. Figure 7-10 is a digital elevation model for Turkey Creek Watershed, Colorado. The digital elevation model is used to derive topographic parameters such as slope and slope length.

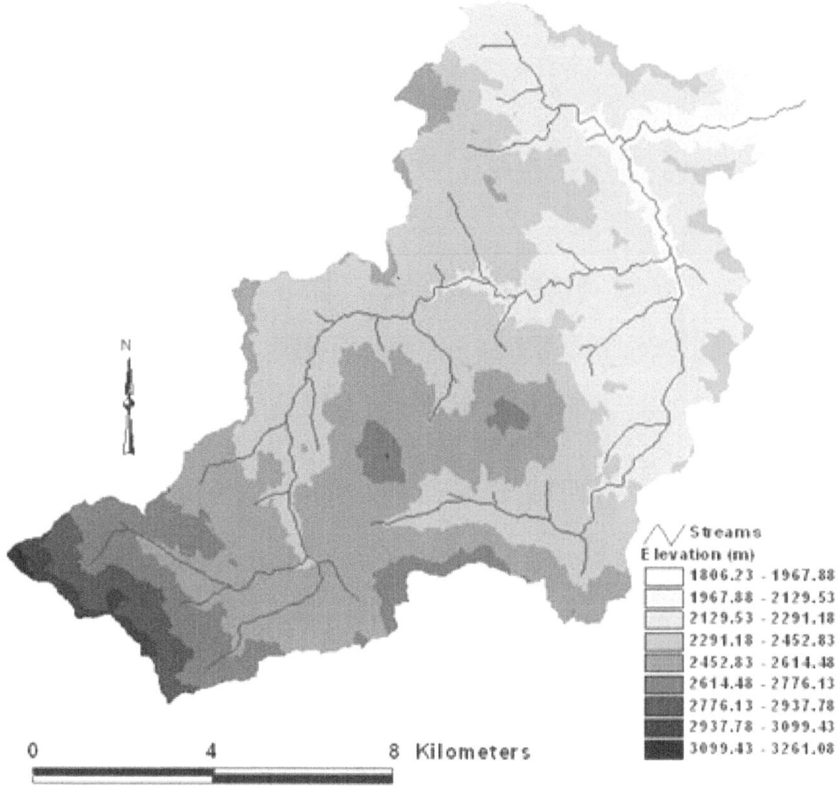

Figure 7-10. Digital Elevation Model for Turkey Creek Watershed, Colorado.
Geza and McCray, 2008. Reprinted with permission from Elsevier.

7.4.4 Climate

Most watershed models generally run on a daily time step and require daily records of precipitation, temperature, cloud cover, dew point temperature, barometric pressure, and wind speed. These data can be obtained from a local source, such as the state-level water resource department. The National Climate Data Center (NCDC) cooperative station data (min/max temp, precipitation only) have higher density of stations and more complete and current data records. Data on wind speed and barometric pressure is often available from the nearest airport meteorological station. NCDC data are available online at http://cdo.ncdc.noaa.gov/CDO/mapproduct (map-based download) and http://cdo.ncdc.noaa.gov/CDO/dataproduct (regular download). Weather Data Management (WDM) data on an hourly time step are also available from BASINS, and can be converted to daily data using a conversion tool in BASINS. Free data for major airports from NCDC containing daily records of precipitation, min/max temperature, wind speed, dew point temperature, cloud cover and air pressure are available at http://www.ncdc.noaa.gov/oa/climate/onlineprod/drought/xmgr.html.

7.4.5 Ponds, Lakes, and Reservoirs

If ponds or reservoirs are in the watershed, they need to be included in the modeling process. The most common data source for defining the locations of waterbodies is the National Hydrography Dataset (NHD). The NHD is a GIS dataset containing surface water features including rivers, streams, lakes, and ponds. Low and medium resolution NHD data are available for the entire conterminous U.S., and high resolution NHD data are available for much of the country. These data are available online from the USGS via http://nhd.usgs.gov/ . The data contains all perennial as well as most intermittent streams, drainage ditches, and other surface water features. The NHD data have recently been converted to a form called NHDinGeo, which is distributed in ESRI Geodatabase format and includes network-tracing capabilities. Lines through waterbodies are inferred so that a continuous stream network is available for the entire U.S. Additional data on lake geometry, such as stage-area and stage discharge (rating curve) for reservoirs, can be obtained from the U.S. Bureau of Reclamation (USBR).

7.4.6 Diversions and Reservoir Release

Water that is managed and controlled using diversions and reservoirs must be included in models that are designed to accurately predict stream flow. These data are very site specific and are not often available from national databases via the Internet. Reservoir release records can be obtained from the stakeholder operating the reservoir (e.g., USBR or utility company). Water diversions for agricultural uses can be obtained from a State Engineer's Water Master. Municipal and industrial (M&I) diversion records can be obtained from the stakeholder that uses the diverted water (such as a city or a power plant).

7.4.7 Air Quality

Some models accept air quality data on concentrations (weekly or monthly) of main constituents in rain water (e.g., WARMF). This can be especially important for nitrogen in areas where deposition is high. Little to no data exists for phosphorus or relevant organics.

The dry deposition data can be obtained from U.S. EPA Clean Air Status and Trends Network (CASTNET), which measures the dry deposition of particles at 55 site locations mostly in the eastern United States. The data are available at http://www.epa.gov/castnet/data.html.

The wet deposition data can be obtained from the National Atmospheric Deposition Program (NADP), which collects data for over 200 sites in the United States, Puerto Rico, and the Virgin Islands. The data are available at http://nadp.sws.uiuc.edu/.

7.5 Measured Output Variables

7.5.1 Stream Flow

Observed stream flow data are required for calibration of the hydrologic component of a model. USGS gage data are typically the most prominent source for surface water observations. The stream flow for many watersheds can be obtained from the U.S. Geological Survey (USGS) National Water Information System (NWIS) at http://waterdata.usgs.gov/nwis/.

The BASINS 3.1 also has a provision to download the stream flow data for each HUC. Historical gage data can also be obtained from state water resources departments, also a public source. The U.S. Bureau of Reclamation (USBR) also operates a network of automated hydrologic and meteorological monitoring stations in many regions. For instance, in the Pacific Northwest, this network and its associated communications called Hydromet (go to http://www.usbr.gov/pn/hydromet/) provide real time and historical data.

Finally, it is possible that counties may have collected stream flow data for certain applications or problems in a watershed. This is rare, but the possible existence of such data should be investigated.

7.5.2 Water Quality/Pollutant Concentrations

For water quality calibration, observed time series data such as concentration of nutrients, heavy metals, and organics (depending on the pollutant of interest) are required. These data are usually limited, and often do not exist in parts of a watershed where OWS are common.

The U.S. Environmental Protection Agency (U.S. EPA) maintains two data management systems containing water quality information for the nation's waters: the Legacy Data Center (LDC) and STORET. The LDC contains historical water quality data dating back to the early part of the 20th century and collected up to the end of 1998. STORET is EPA's main repository of water quality monitoring data for surface water and groundwater. STORET contains data collected beginning in 1999, along with data that was documented and migrated from the LDC. Both systems contain raw biological, chemical, and physical data on surface and groundwater collected by federal, state and local agencies, academics, and others. STORET data can be obtained directly from EPA at http://www.epa.gov/storet/.

The BASINS framework on the U.S. EPA web page for watershed models also provides a tool to download USGS water quality data automatically for each HUC. This may be readily used with watershed models WARMF and SWAT.

Additional data for streams, lakes, springs, and groundwater can also be found from the USGS website (http://waterdata.usgs.gov/nwis). These pages provide access to water-resources data collected at approximately 1.5 million sites in all 50 states, the District of Columbia, and Puerto Rico. The USGS investigates the occurrence, quantity, quality, distribution, and movement of surface and underground waters and disseminates the data to the public, state and local governments, public and private utilities, and other federal agencies involved with managing water resources.

Data are also likely to be available from state environmental and health agencies via the internet. State departments should be contacted to determine the availability of data for the watershed of interest.

County departments of health, environment, planning, etc., may also sometime stream or groundwater data. These data are usually found in hard copy or PDF reports, and are not usually available on line.

In many watersheds, sampling programs are conducted by various stakeholder groups. It is advisable to look for such sources. For example, in Oregon's Deschutes River watershed, the Deschutes Watershed Council conducted several stream water quality sampling campaigns. In Colorado's Turkey Creek watershed, a conglomeration of stakeholders – including residents, developers, county planning administrators, and the USGS – conducted a mountain groundwater study.

Finally, it is possible, even likely, that a university or other independent research agency has conducted a study that could provide multiple years of data for streams, lakes, or groundwater. These are often published in the peer-reviewed literature, and can often be located by online search engines like Google Scholar.

7.6 Summary Tables

Model category and data requirements/inputs are listed in Table 7-4. Table 7-5 is a summary of data sources for different types of models, public and local sources of the input data.

Table 7-4. Model Types and Data Needs.

Model type	Data inputs	Data source
Mass balance screening	Human population; Number of septic systems Surface water discharge; Groundwater outflow/recharge Measured nutrient concentration	Table 7-5
GIS-based screening	GIS data layers of elevation (DEM), soils, and land use ; Precipitation; Evapotranspiration; Surface water discharge Groundwater outflow/recharge	Table 7-5
Surface water	GIS data layers of elevation (DEM), soils, and land use are used to generate the input files Precipitation, minimum and maximum temperature, dew point, air pressure, wind speed Point sources (wastewater treatment plants, industries) Air Quality (dry and wet deposition) Regulated flow (reservoir release, diversions) Observed hydrology data (stream flow, reservoir elevation)	Table 7-5
Groundwater	Observation head; Hydraulic conductivity; Transmissivity Boundary conditions	Table 7-5
Integrated watershed	Some or all of surface and groundwater data listed above may be required DEM, land cover data, soils data, reservoir release data, daily precipitation and temperature Point source data (industries, wastewater), geologic data, hydraulic conductivity, groundwater Surface water interaction (Leakage coefficients and the magnitudes of gaining and losing reaches) Groundwater withdrawals, and boundary conditions	Table 7-5

Table 7-5. Data Type and Sources.

Dataset	Source Agency / Company	Source Address / Location
Digital Elevation Model (DEM)	USGS MAPMART Other Sources	http://seamless.usgs.gov http://www.mapmart.com/DEM/DEM.htm
National Hydrography Dataset (NHD)	USGS	http://nhd.usgs.gov/
National Land Cover Data (NLCD) EPA GIRAS Land Use Land Cover (LULC)	USGS EPA	http://seamless.usgs.gov EPA BASINS Model
STATSGO SSURGO	NRCS NRCS	http://www.ncgc.nrcs.usda.gov/products/datasets/statsgo/ http://www.ncgc.nrcs.usda.gov/products/datasets/ssurgo/index.html
Climatic data	National Climate Data Center (NCDC)	NCDC: Global Summary of the Day http://www.ncdc.noaa.gov/oa/climate/onlineprod/drought/xmgr.htmlNC DC: Cooperative Station Database http://cdo.ncdc.noaa.gov/pls/plclimprod/poemain.accessrouter?datasetabbv=SOD NRCS SNOTEL Data (Western U.S.) http://www.wcc.nrcs.usda.gov/snotel/
Evapotranspiration		Local sources, ET models
Point Sources	EPA Pollution Control System (PCS) /NPDES	Monthly data for major dischargers http://www.epa.gov/enviro/html/pcs/pcs_query_java.html
Agricultural management		Data: fertilizer application rates, tillage types, harvesting operations
Air Quality (wet deposition) Air Quality (dry deposition)	National Atmospheric Deposition Program (NADP) Clean Air Status and Trends Network (CASTNET)	Weekly precipitation data for major cations and anions http://nadp.sws.uiuc.edu/sites/ntnmap.asp? Air quality and deposition velocities http://www.epa.gov/castnet/data.html
Regulated flow (Reservoir Release)	USGS gage downstream of dams	http://waterdata.usgs.gov/nwis/sw Data: Diversions, Reservoir Releases
Observed Hydrology Data	USGS USBR	Discharge http://waterdata.usgs.gov/nwis/swDischarge http://www.usbr.gov/pn/hydromet/
Observed Water Quality Data	USGS EPA (STORET) Other Sources	http://nwis.waterdata.usgs.gov/usa/nwis/qwdata http://www.epa.gov/storet/dbtop.htmlLocal sources (Water districts, Dischargers, State and city)
Groundwater data Observation head Hydraulic conductivity Transmissivity Boundary conditions	Local data	Observation wells Pump test/slug test Driller's report Mass balance models (precipitation minus evapotranspiration)

CHAPTER 8.0

IDENTIFYING INPUT PARAMETER VALUES FOR WATERSHED MODELS

8.1 Introduction

Numerous parameters are required for modeling heterogeneous watersheds, especially under time-varying conditions. Model parameters can be measured, estimated via calibration, or estimated independently outside of calibration. Some parameter values in watershed modeling can be directly obtained from field, laboratory or from previous research reports.

The first step in determining model input parameters is to identify parameters that can be reasonably measured. Alternatively, if measured values are not available, calibration can be used to estimate optimal values that provide the best match between field observations and their equivalent simulated values. Parameters can also be estimated independent of calibration using information on watershed characteristics. These parameters are not directly measured but can be computed or estimated from measured data.

Most watershed models are divided into smaller catchments and river segments to increase discretization and reduce the effect of parameter lumping. Each of these catchments and river segments are represented by a number of parameters (river and catchment parameters). There are also parameters that apply uniformly to all catchments in the watershed. The number of model parameters could be large if different parameters are estimated via calibration for each catchment or river segment.

For models with lengthy execution times, simulation runs can be time consuming if many parameters are estimated via calibration. In addition to length of simulation time, in most situations, few observations are available, and thus they support estimation of only a few model input parameters via calibration. Use of many parameters can lead to a better match between simulated and observed values. However, at times the improved fit is capturing errors in the observations rather than behavior of the system and it is often impossible to converge on a unique solution when estimating many parameters. It also increases the uncertainty associated with the estimated input values and the resulting predictions.

Models that have too many input parameters in relation to the available field data are likely to not be unique. Thus, input parameter values can be correlated and not unique, or the model output is not sensitive to a wide range of values for certain input parameters. Multiple combinations of parameter values may result in equally good fits to observed data, but such non-unique models have little chance of providing good predictions in the future (for time periods outside that used for data collection or for scenarios that differ from calibration conditions) because values for many of the model parameters may not represent actual watershed hydrologic conditions.

It is usually necessary to reduce the number of independent model-input parameters for a watershed-scale model. Generally, simplification of the model can reduce the number of values to be estimated. Thus, most of the parameters have to be either measured or estimated.

There are some parameters that cannot be directly measured (e.g., curve number used to compute surface runoff in SCS method). Such parameters have to be either estimated from measured data or determined via calibration. Other parameters are more appropriately measured than determined through calibration. Whether a parameter is measured or estimated may also depend on the size of the watershed. For a field scale model, the slope or slope length can be measured while for a bigger watershed it may be estimated from a DEM using GIS. Some parameters can fall in either category (measured, estimated or calibrated). So, for any model and for some parameters there is no distinct category.

Selection of models that can be applied to OWS requires evaluation of key features of the models and ability to handle non-point source pollution from OWS. Emphasis has to be given to models that can simulate the most common wastewater pollutants, especially nitrogen and phosphates both at a field and watershed scale. Sediment transport should also be considered since transport and fate of sediments and nutrients are intimately related.

The models included in the initial review include AGNPS, ANNAGNPS, ANSWERS-2000, CREAMS-WT, GLEAMS, HSPF, MIKE-SHE, SWAT, MODFLOW, SWMM, WARMF, WMS and GIS Screening Models.

In general, models such AGNPS, ANSWERS-2000, ANNAGNPS, CREAMS-Wt, GLEAMS, SWAT and HSPF have similar routines for nutrient transformation. Nitrogen, phosphorus, and pesticides in these models are based on routines developed for the CREAMS/GLEAMS models including biochemical processes and groundwater loading. None of these models explicitly account for OWS.

AGNPS and ANSWERS-2000 are primarily surface runoff models and do not handle subsurface flow well, and therefore are not suited to simulate OWS pollutants.

Models like ANNAGNPS, CREAMS-WT and GLEAMS have routines to simulate subsurface flow and leaching but subsurface flow and leaching do not contribute to stream flows. Although these models can be good to simulate the effect of OWS pollutants on groundwater, they are not suited to simulate the impact of OWS pollutants on stream flow.

On the other hand, models like SWAT and HSPF have routines to simulate lateral and groundwater contributions to stream flow. Although these models don't explicitly account for OWS pollutants, they can still be used by applying the non-point and point source routines of the models. WARMF has a routine that accounts for OWS. Compared to WARMF, HSPF and SWAT model do not explicitly account for OWS. Application of SWAT model for fate and transport of nitrogen derived from OWS thus far was based on the fertilizer input routine in SWAT. Research is currently underway to incorporate OWS routine in SWAT.

This chapter will focus on SWAT and WARMF parameters. Additional information on capabilities of various watershed scale models is provided in Appendix C in the "review of distributed models used for watershed-scale assessments of onsite wastewater system impacts."

8.2 Method of Obtaining Parameter Values

8.2.1 Measured Parameters

It is important to make the best use of measured, available data or known conditions in the watershed to reduce the number of calibrated parameters. The first step to reduce the number of parameters to be estimated via calibration is to identify model parameters that could be determined with reasonable accuracy from field, laboratory or previous research reports. Measured parameters may include topography, soil, and land cover, weather parameters that are directly measured and used as input in the model. Measured values are available throughout a watershed for certain catchment specific parameters, such as slope, aspect, and elevation. These are simply input to the model, reducing the number of parameters to be estimated via calibration.

8.2.2 Estimated Parameters (By Calibration)

Some parameters cannot be measured directly; therefore, they must be determined through calibration. This is done iteratively through calibration to best match measured data. As mentioned earlier, the number of parameters estimated via calibration has to be reduced. For those parameters that are not expected to vary much from catchment to catchment, the same parameter values may be used in every catchment. For others, parameter values for one spatial unit can be derived from parameters values from other spatial units in the watershed. A relationship between required parameters of one spatial unit and corresponding parameters of another unit can be developed using available information about the parameters. The methods discussed (identifying parameters that could be measured, estimating parameters from other measured values, using identical values for some parameters and deriving parameter for one spatial unit or depth from another spatial unit or depth) are some of the approaches to reduce the number of parameters that need to be determined via calibration.

Even after using these techniques, the number of parameters to be determined via calibration could still be large. The number of parameters can be further reduced by investigating the sensitivities. Model outputs are not equally sensitive to all parameters of a model. If an output is not sensitive to a parameter, it is advisable to assign estimates for those parameters and consider only the parameters to which the model is sensitive during the calibration.

8.2.3 Estimated Parameters (Outside of Calibration)

Not all parameters can be directly measured or determined through calibration. Some model parameters have to be determined from watershed characteristics such as land use or soil type or a unique combination of watershed characteristics. Sometimes, a model input parameter that is directly measured for one model type, can be an estimated parameter for another model type. In a small field scale model, for instance, the slope and slope length can be measured.

For a watershed scale model, these parameters can be derived or estimated from a DEM, and most of the inputs are estimated from other inputs. A DEM is used to calculate parameters such as slope and slope length as well as the definition of stream network. Soil parameters can be estimated from soil data according to the State Soil Geographic (STATSGO) or Soil Survey Geographic Database (SSURGO). Curve numbers used for runoff computation are estimated as a function of hydrologic soil group, land use land cover type, and antecedent moisture condition.

Another example is the estimation of evapotranspiration in WARMF. Hargreaves (1974) equation calculates monthly potential ET based on temperature, relative humidity and a parameter that depends on geographic location. WARMF estimates the daily PET using the monthly PET and a calibration parameter called evaporation magnitude. So some parameters can be estimated as a function of measurable quantities.

Those model parameters that could be estimated with reasonable accuracy based on measured data should be identified through initial screening. Besides reducing the number of parameters to be estimated via calibration, it may sometimes be more certain to estimate parameter values from field data than estimating them via calibration.

8.2.4 Parameterization of SWAT and WARMF

Selection of models that can be applied to OWS requires evaluation of key features of the models and ability to handle non-point source pollution from OWS. Emphasis has to be given to models that can simulate the most common wastewater pollutants, especially nitrogen and phosphates both at a field and watershed scale. Sediment transport should also be considered since transport and fate of sediments and nutrients are intimately related.

Models such as SWAT and HSPF have routines to simulate lateral and groundwater contributions to stream flow. Although these models don't explicitly account for OWS pollutants, they can still be used by applying the non-point and point source routines of the models. Compared to WARMF, HSPF and SWAT model do not explicitly account for OWS. SWAT and WARMF have been used to some extent to evaluate the impact of OWS pollutants. Pradhan et al. (2005) used the SWAT model to evaluate fate and transport of nitrogen derived from OWS using the fertilizer input routine in SWAT. Weintraub et al. (2004) used WARMF model as a tool for tracking fate and transport of nutrients from OWS. Watershed modeling using the SWAT model was performed to understand the potential influence of various point and non-point sources of P in the Blue River watershed in Colorado (Lemonds and McCray, 2003).

This chapter presents parameters that are used in SWAT and WARMF.

Table 8-1 and Table 8-10 list parameters used to calibrate WARMF and SWAT models for hydrology, sediment, phosphorus, nitrogen and pesticides and organics. The tables also give the range of values for each parameter. The ranges are based on WARMF (Herr et al., 2001) and SWAT literature (Neitsch et al., 2004, Neitsch et al., 2005) and indicate if the parameters can be most preferably measured (M), estimated (E) or calibrated (C). Some parameters can fall in either category (measured, estimated or calibrated). So for any model and for some parameters, there is no distinct category.

For WARMF model, the tables also indicate whether a parameter applies to an entire basin (system coefficients), a catchment, or a stream segment.

The SWAT model defines parameters at the basin/watershed, subbasin and hydrologic response unit (HRU) levels. The first level of watershed subdivision in SWAT is the sub-basin. Sub-basins possess a geographic position in the watershed and are spatially related to one another. A sub-basin contains one or more HRUs, a tributary channel and a main channel or reach. HRUs are portions of a sub-basin that possess unique land use/management/soil attributes.

The .bsn files in the subsequent tables are for parameters defined at the basin/watershed level. The .rte files are for parameters used to define the stream segment (routing parameters) at the subbasin level. The remaining parameters (in files hru, sol, chm, gw, mgt) are defined at the smallest subdivision or HRU level.

SWAT uses five databases to store information required for plant growth, urban land characteristics, tillage, fertilizer components and pesticide properties:

- Information required to simulate plant growth is stored by plant species in the plant growth database file (crop.dat).

- The urban database file (urban.dat) summarizes urban landscape attributes needed to model urban areas.
- The tillage database (till.dat) contains mixing depth and mixing efficiency data for the most common tillage implements. Tillage operations redistribute nutrients, pesticide and residue in the soil profile.
- The fertilizer database (fert.dat) summarizes the relative fractions of nitrogen and phosphorus pools in the different fertilizers.
- The pesticide database file (pest.dat) summarizes pesticide attribute information for various pesticides.

Some of the parameters in crop data base files can be adjusted during calibration.

These models (WARMF and SWAT) are highly parameterized and not all the parameters are listed. The list includes most relevant parameters based on user manuals and the literature describing calibration and sensitivity analyses involving these models. Information on sensitivity of these parameters is useful to model users. Not all the hydrology and water quality parameters in the list are sensitive parameters. A sensitivity analysis of the WARMF parameters for hydrology and water quality is given in case studies (Appendix B). The SWAT parameters listed in subsequent tables are those frequently used in calibration. A list of literature on sensitivity for the SWAT parameters is available at (http://www.brc.tamus.edu/swat/swat-peer-reviewed_categorized.pdf). This can be used as a guide to deciding which parameters should be adjusted during calibration and which parameters should be simply set to a reasonable value depending on the watershed attributes.

Table 8-1. WARMF Model Adjustable Hydrologic Input Parameters.

Parameter	Range of values	M/E/C	Units
System Coefficients			
- Land use category			
Fraction impervious (Residential)	0-1	E/C	--
Fraction impervious (Commercial)	0-1	E/C	--
- Physical data category			
Evaporation magnitude	0.6-1.4	E/C	deg c
Evaporation skewness	0.7-1.2	E/C	deg c
- Snow category			
Snow formation temperature	0-3	E/C	cm/°c/day
Snowmelt temperature	0-3	E/C	cm/°c/day
Snowmelt rates (open & forest areas)	0.05-0.10	E/C	cm/day
Sublimation rates (open & forest area)	0.005	E/C	cm/day
Catchments coefficients			
- Physical data category			
Detention storage	0-100	E/C	%
Roughness	0.01-0.4	E/C	--

- Meteorology category			
Precipitation weighting factor	0.5-1.5	M/E/C	--
Temperature lapse rate	-5-5	M/E/C	deg c
Altitude lapse rate	0.001-0.009	M/E/C	deg c/m
- Soil coefficients category			
Initial moisture	0-0.6	M/E/C	m^3/m^3
Field capacity	0.0-0.4	M/E/C	m^3/m^3
Saturation moisture	0.2-0.6	M/E/C	m^3/m^3
Horizontal conductivity	>0	M/E/C	m/day
Vertical conductivity	>0	M/E/C	m/day
River Coefficients			
- Physical data category			
Manning's n	0.02-.045	E/C	--

M=preferably measured, E=preferably Estimated, C=preferably calibrated

Table 8-2.SWAT Model Adjustable Hydrologic Input Parameters.

Parameter	Range of values	M/E/C	Units	File type
Surface water				
Maximum melt rate for snow during the year (SMFMX)	0-10	E/C	mm/day-o_C	.bsn
Minimum melt rate for snow during the year (SMFMN)	0-10	E/C	mm/day-o_C	.bsn
Snow pack temperature lag factor (TIMP)	0-1	E/C	--	.bsn
Snowmelt base temperature (SMTMP)	-5-5	E/C	deg c	.bsn
Threshold depth of snow above which there is 100% cover (SNOWCOVMX)	0-500	E/C	mm	.bsn
Fraction of SNOWCOVMX at 50% snow (SNO50COV)	0-1	E/C	--	.bsn
Snowfall temperature (SFTMP)	-5-5	E/C	deg c	.bsn
Surface runoff lag time (SURLAG)	1-24	E/C	days	.bsn
Muskingum routing parameter (MSK_X)	0-0.3	E/C	--	.bsn
Muskingum routing coefficient (MSK_CO1/ MSK_CO1)	0-10	E/C	--	.bsn
Reach evaporation adjustment factor (EVRCH)	0.5-1	E/C	--	.bsn
Soil evaporation compensation factor (ESCO)	0-1	E/C	--	.hru
Plant uptake compensation factor (EPCO)	0-1	E/C	--	.hru
SCS CN for moisture condition II (CN2P)	25-98	E/C	--	.hru
Available water capacity of the soil layer (SOL_AWC)	0-1	M/E/C	%	.sol
Saturated hydraulic conductivity (SOL_K)	0-2000	M/E/C	mm/hr	.sol
Maximum crack volume of the soil profile (SOL_CRK)	0-1	E/C	m3/m3	.sol
Manning's n value for pasture (OV_NP)	0.01-30	E/C	--	.hru
Manning's n value for forest (OV_NF)	0.01-30	E/C	--	.hru
Lateral flow travel time (LAT_TTIME)	0-180	E/C	days	.hru
Maximum canopy storage for pasture (CANMXP)	0-100	E/C	mm	.hru
Maximum canopy storage for forest land (CANMXF)	0-100	E/C	mm	.hru
Groundwater				
Base flow alpha factor (ALPHA_BF)	0-1	E/C	days	.gw
Groundwater revap coefficient (GW_REVAP)	0.02-0.2	E/C	mm	.gw
Threshold depth of water in the shallow aquifer for 'revap' to occur (REVAPMN)	0-500	E/C	mm	.gw
Groundwater delay (GW_DELAY)	0-500	E/C	days	.gw
Deep aquifer percolation fraction (RCHRG_DP)	0-1	E/C	--	.gw
Threshold depth of water in the shallow aquifer for return flow to occur (GWQMIN)	0-5000	E/C	mm	.gw

Table 8-3. WARMF Model Adjustable Sediment Input Parameters.

Parameter	Range of values	M/E/C	Units
System Coefficients			
- Land use category			
Rainfall Detachment Factor	$1.1e^{-4}$-$1.1e^{-1}$	E/C	mg/g dry wt
Flow Detachment Factor	0.9	E/C	g/cm^3
Cropping factor	0.01-1.0	E/C	g/cm^3
- Sediment category			
Sediment size	0.002-2	M/E	mm
Specific gravity	2.65	M/E	None
Settling velocity	0.004-12	M/E	mm/s
Catchments coefficients			
- Sediment category			
Soil Erosivity Factor	0.2-0.4	E/C	None
River Coefficients			
- Sediment category			
Initial depth of sediment	>=0	E/C	m
Detachment velocity multiplier	$\sim 1.0e^{-6}$	E/C	Empirical
Detachment velocity exponent	1.0 - 2.0	E/C	None
Vegetation factor	> 0	E/C	None
Bank stability factor	> 0	E/C	None
Bed diffusion rate	> 0	E/C	m2/d
- Reactions Category			
Sediment settling	> =0	M/E	m/d

Table 8-4. SWAT Model Adjustable Sediment Input Parameters.

Parameter	Range of values	M/E/C	Units	File type
Linear parameter for channel sediment routing (SPCON)	0.001-0.01	E/C	--	.bsn
Exponent parameter for channel sediment routing (SPEXP)	1.0-1.5	E/C	--	.bsn
Adjustment factor for sediment routing- tributary channels (APM)	0.5-2.0	E/C	--	.bsn
Adjustment factor for sediment routing- main channel (PRF)	0.0-2.0	E/C	--	.bsn
USLE soil erodibility factor (USLE_K)	0.0-0.65	E/C	--	.sol
Minimum value of MUSLE C factor (USLE_C)	0.001-0.5	E/C	--	crop.dat
Support practice factor (USLE_P)	0.1-1.0	E/C	--	.mgt
Slope length (SLSUBBSN)	10-150	E/C	m	.hru
Average slope (HRU_SLP)	0.0-0.6	E/C	%	.hru
Percent rock in the first soil layer (ROCK)	0-100	E/C	%	.sol
Channel cover factor K (CH_COV)	-0.001-1.0	E/C	--	.rte
Channel erodibility factor K (CH_EROD)	-0.05-0.6	E/C	--	.rte

Table 8-5. WARMF Model Adjustable Phosphorus Input Parameters.

Parameter	Range of values	M/E/C	Unit
System Coefficients			
- Land use category			
Leaf composition	0-20	E/C	mg/g dry wt
Trunk composition	0-1	E/C	mg/g dry wt
- Septic Category			
Septic system discharge	0.24-15	M/E	mg/l
- Litter Category			
Litter decay rate	1-2	E/C	$year^{-1}$
Fine litter decay rate	0.06	E/C	$year^{-1}$
Humus decay rate	0.005	E/C	$year^{-1}$
Catchments coefficients			
- Soil coefficients category	0-40	E/C	meq/100g
Cation Exchange Capacity	400	E/C	mg PO4/kg
Maximum phosphate ads.	0-500	E/C	L/KG
Adsorption isotherm	>=0	M/E/C	mg/l
Initial conc. of phosphate	>=0	M/E	kg/ha P
- Land application category			
Phosphate	0.2-1.0	E/C	day^{-1}
River Coefficients			
- Reactions Category			
Aeration Factor	0.1-2	E/C	g / m2 / day
Sediment Oxygen Demand	0-1	E/C	m/d
Precipitate settling	> 0	E/C	day^{-1}
BOD Decay	0.1-1.0	E/C	day^{-1}
Organic Carbon decay	> =0	M/E/C	mg/l P
Initial concentration of phosphate			
- Adsorption Category			
Adsorption isotherms	> =0	E/C	L/kg

Table 8-6. SWAT Model Adjustable Phosphorus Input Parameters.

Parameter	Range of values	M/E/C	Units	File type
Phosphorus percolation coefficient (PPERCO)	10-17.5	E/C	10m3/kg	.bsn
Phosphorus soil partitioning coefficient (PHOSKD)	100-200	E/C	10m3/kg	.bsn
Phosphorus availability index (PSP)	0.01-0.7	E/C	--	.bsn
Residue decomposition factor (RSDCO)	0.02-0.1	E/C	--	.bsn
Phosphorus uptake distribution parameter (UBP)	0-100	E/C	--	.bsn
Initial organic P in soil the layer (SOL_ORGP)	0-4000	E/C	ppm	.chm
Initial soluble P conc. in the soil layer (SOL_SOLP)	0-100	E/C	ppm	.chm
Phosphorus enrichment ratio (ERORGP)	0-5	E/C	--	.hru
Biological mixing efficiency (BIOMIX)	0-1	E/C	--	.mgt
Fraction of P in plant at emergence (PLTPFR 1)	0.0005-0.01	E/C	--	crop.dat
Fraction of P in plant at 50% maturity (PLTPFR 2)	0.0002-0.007	E/C	--	crop.dat
Fraction of P in plant at full maturity (PLTPFR 3)	0.0003-0.004	E/C	--	crop.dat
Fraction of phosphorus in the yield (CPYLD)	0.0001-0.015	E/C	--	crop.dat
Fraction of mineral P in the Fertilizer (FMINP)	0-1	E/C	--	fert.dat
Fraction of organic P in the Fertilizer (FORGP)	0-1	E/C	--	fert.dat
Soluble P conc. in the groundwater flow (GWSOLP)	0-1000	E/C	mg/l	.gw
Mixing efficiency for tillage operation (EFFMIX)	0-1	E/C	--	till.dat
Depth of mixing by tillage operation (DEPTIL)	0-750	E/C	mm	till.dat

Table 8-7. WARMF Model Adjustable Nitrogen Input Parameters.

Parameter	Range of values	M/E/C	Units
System Coefficients			
- Land use category			
Leaf composition	0-20	E/C	mg/g dry wt
Trunk composition	0-1	E/C	g/cm^3
Litter fall rate	0-0.16	E/C	kg/m^2/month
Productivity	0-0.8	E/C	kg/m^2/year
- Septic category			
Septic system discharge quality (NH4)	17-178	M/E/C	mg/l
Septic system discharge quality (NO3)	0-1.94	M/E/C	mg/l
- Litter Category			
Litter decay rate	1-2	E/C	$year^{-1}$
Fine litter decay rate	0.06	E/C	$year^{-1}$
Humus decay rate	0.005	E/C	$year^{-1}$
Litter leachable ions	0-0.2	E/C	None
Fine litter leachable ions	0.2-0.7	E/C	None
Humus leachable ions	0.2-0.7	E/C	None
Non-structural leachable ions	0.05	E/C	None
- Canopy category			
Foliar nitrification	0.4	E/C	day^{-1}
Gas deposition	0.2-0.8	E/C	cm/sec
Gas uptake	0.1	E/C	cm/sec
Catchments coefficients			
- Adsorption category			
Cation Exchange Capacity	>=0	E/C	meq/100g
Initial conc. of nitrate/ammonia	>=0	E/C	mg/l
Initial base saturation (NH4 and H)	>=0	E/C	%
- Reactions Category			
Nitrification and denitrification rates	>=0	E/C	day^{-1}
- Land application category			
Nitrate/Ammonia	>=0	M/E	kg/ha N
River Coefficients			
- Reactions Category			
Aeration Factor	0.2-1.0	E/C	day^{-1}
Sediment Oxygen Demand	0.1-2	E/C	g/m2/day
Precipitate settling	0-1	E/C	m/d
BOD Decay	> 0	E/C	day^{-1}
Organic carbon decay	0.1-1.0	E/C	day^{-1}
Initial conc. of nitrate/ ammonia	> =0	E/C	mg/l N
Nitrification and denitrification rates (water and bed)	> =0	E/C	day^{-1}
- Adsorption Category			
Initial base saturation (NH4)	> =0	E/C	L/kg

Table 8-8. SWAT Model Adjustable Nitrogen Input Parameters.

Parameter	Range of values	M/E/C	Units	File type
Concentration of Nitrogen in the Rain (RCN)	0-15	E/C	mg N/l	.bsn
Rate coefficient for mineralization of the humus active organic nutrients (CMN)	0.001-0.003	E/C	--	.bsn
Rate coefficient for mineralization of the residue fresh organic nutrients (RSDCO)	0.02-0.1	E/C	--	.bsn
Rate coefficient for denitrification (CDN)	0.02-0.25	E/C	--	.bsn
Denitrification Threshold water content (SDNCO)	~1.0	E/C	--	.bsn
Nitrate percolation coefficient (NPERCO)	0-1	E/C	--	.bsn
Nitrate uptake distribution parameter (UBN)	0-1	E/C	--	.bsn
Fraction of porosity from which anions are excluded (ANION_EXCL)	0.01-1.0	E/C	ppm	.sol
Initial organic N in soil the layer (SOL_ORGN)	0-10000	E/C	ppm	.chm
Initial nitrate conc. in the soil layer (SOL_NO3)	0-100	E/C	--	.chm
Organic nitrogen enrichment ratio (ERORGN)	0-5	E/C	--	.hru
Biological mixing efficiency (BIOMIX)	0-1	E/C	--	.mgt
Fraction of N in plant at emergence (PLTNFR 1)	0.004-0.07	E/C	--	crop.dat
Fraction of N in plant at 50% maturity (PLTNFR 2)	0.002-0.05	E/C	--	crop.dat
Fraction of N in plant at full maturity (PLTNFR 3)	0.001-0.27	E/C	--	crop.dat
Fraction of nitrogen in the yield (CNYLD)	0.0015-0.075	E/C		crop.dat
Fraction of mineral N in the Fertilizer (FMINN)	0-1	E/C	--	fert.dat
Fraction of organic N in the Fertilizer (FORGN)	0-1	E/C	--	fert.dat
Fraction of ammonium N in the Fertilizer (FNH3N)	0-1	E/C	--	fert.dat
Nitrate conc. in the groundwater flow (GWNO3)	0-1000	E/C	mg/l	.gw
Mixing efficiency for tillage operation (EFFMIX)	0-1	E/C	--	till.dat
Depth of mixing by tillage operation (DEPTIL)	0-750	E/C	mm	till.dat

Table 8-9. WARMF Model Adjustable Pesticide/Organics Input Parameters.

Parameter	Range of values	M/E/C	Units
System coefficients			
- Septic category			
STE concentration (organic carbon)	--	M/E	Mg/l
Catchments coefficients			
- Soil Category			
Adsorption isotherm (pesticides)	>=0	E/C	L/kg
Adsorption isotherm (dissolved organic carbon)	>=0	E/C	L/kg
Initial concentration (pesticide)	>=0	M/E/C	mg/l
Initial concentration (dissolved organic carbon)	>=0	M/E/C	mg/l
- Reactions category			
Decay rates (pesticide)	0.1-1.0	E/C	day^{-1}
Decay rates (organic carbon)	0.1-1.0	E/C	day^{-1}
- Land application category			
Land application (pesticide)	>=0	M/E	kg/ha
Land application (organic carbon)	>=0	M/E	kg/ha
River Coefficients			
- Adsorption Category			
Adsorption isotherm bed/water (pesticides)	>=0	E/C	L/kg
Adsorption isotherm bed/water (organic carbon)	>=0	E/C	L/kg
- Reactions category			
Decay rates (pesticide)	0.1-1.0	E/C	day^{-1}
Decay rates (organic carbon)	0.1-1.0	E/C	day^{-1}
- Initial concentration categroy			
Initial concentration (pesticide)	>=0	M/E	mg/l
Initial concentration (organic carbon)	>=0	M/E	mg/l

Table 8-10. SWAT Model Adjustable Pesticide/Organics Input Parameters.

Parameter	Range of values	M/E/C	Units	File type
Pesticide percolation coefficient (PERCOP)	0-1	E/C	--	.bsn
Wash-off fraction for the pesticide (WOF)	0-1	E/C	--	pest.dat
Soil adsorption coefficient normalized for soil organic carbon content (SKOC)	$1\text{-}10^8$	E/C	mg/kg	pest.dat
Pesticide enrichment ratio (PSTENR)	0-5	E/C	--	.chm
Amount of organic carbon in the soil layer (SOL_CBN)	0.05-10	E/C	% soil wt	.sol
Initial amount of pesticide in soil layer (SOLPST)	0-500	E/C	mg/kg	.chm
Initial amount of pesticide on foliage (PLPST)	0-500	E/C	kg/ha	.chm
Solubility of the pesticide in water WSOL	$1\text{-}10^7$	E/C	mg/l	pest.dat
Half-life of the pesticide in the soil HLIFE_S	0-1000	E/C	days	pest.dat
Half-life of the pesticide on foliage (days) HLIFE_F	0-1000	E/C	days	pest.dat
Pesticide application efficiency (AP_EF)	0-1	E/C	--	pest.dat
Mixing efficiency for tillage operation (EFFMIX)	0-1	E/C	--	till.dat
Depth of mixing by tillage operation (DEPTIL)	0-750	E/C	mm	till.dat

For phosphorus, pesticides and organics, calibrating sediment loading and concentration is important.

CHAPTER 9.0

MODEL CALIBRATION, PERFORMANCE EVALUATION, AND LONG-TERM USE

9.1 Introduction

Watershed models are useful tools for simulating the effect of watershed processes and management on soil and water resources. They should pass through calibration and evaluation processes before they are used as a decision making tool. The important steps include data collection, model setup, parameter identification through prior knowledge and sensitivity, calibration, validation, model evaluation, post-processing and evaluation of prediction uncertainty. Each of the steps is presented in the subsequent sections. Model types have variations of the specific physical, chemical, and biological systems they each attempt to represent, so specific application steps or procedures may differ. In general, however, they have many steps in common.

Figure 9-1 shows the general steps in the modeling process. The calibration and validation steps are especially critical since the outcome establishes how well the model represents the watershed for the purpose of the study, although the other steps are also an important part of the process.

Calibration refers to an iterative procedure of parameter evaluation and refinement, as a result of comparing simulated and observed values. Proper model calibration is important in hydrologic modeling studies to reduce uncertainty in model simulations (Engel et al., 2007). Not all model parameters are used in calibration. Some sensitive or insensitive parameters with well known values can be set to a value based on prior knowledge of the watershed characteristics (a priori approach). The sensitivity and correlation analysis can then be applied to identify parameters that can be determined via calibration and those that should again be set to a value. Model validation ensures that the calibrated model properly assesses all the variables and conditions which can affect model results, and demonstrates the ability to predict field observations for periods different from the calibration conditions. The validity of the calibrated model is measured using performance criteria.

There are several model performance criteria, including statistical techniques (e.g., standard regression) and graphical techniques (e.g., visual comparison of simulated and measured constituents). Once the model passes through the various stages it is ready for ultimate use as a decision support tool for management and regulatory purposes.

9.2 Data Collection and Model Setup

The main goal in watershed modeling is to represent the natural system as accurately as possible using data. Thus, watershed models need a diversity of information (land use, soil, geological, topographic, meteorological and hydrological) in order to set up and run. The amount of data required depends on the complexity of the model. Some models can be run using data available in tabular form and may not contain specific spatial or geographical locations. Other models are built from a combination of tabular information and geographically referenced data (i.e., GIS data) or entirely within a GIS from datasets that are geographically referenced such as digital elevation models (DEM), land use-land cover data and soils data. Types of model and data requirement and possible sources were presented in Chapter 7.0.

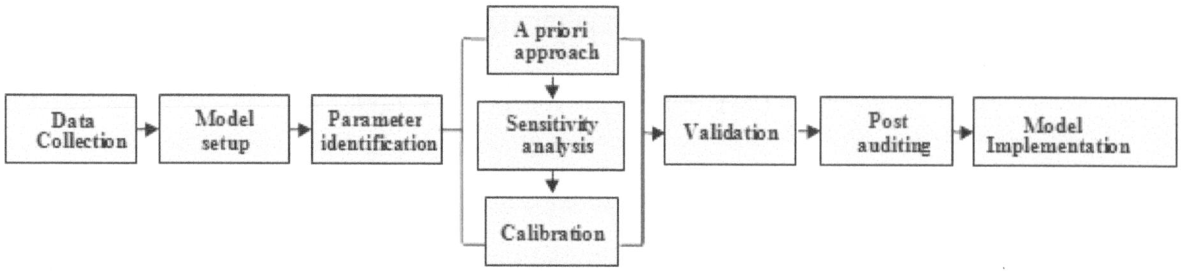

Figure 9-1. The Modeling Process.

9.3 Model Calibration

Watershed models simplify field conditions, and the underlying assumptions and inputs influence the model results.

Model calibration is the process of estimating model parameters by comparing model predictions (output) for a given set of assumed conditions with observed data for the same conditions. Model calibration improves parameter estimates before the model is used for decision making. It is a complex process because model errors can originate from diverse sources, including the limitation in the mathematical structure of the model, limitations of the input and output data, and lack of knowledge of the watershed characteristics.

Calibration requires that we decide which calibration criteria or objective function to use. We must determine the set of parameters that can meet our calibration criteria. There are various methods to determine these set of parameters, ranging from manual or trial-and-error methods to automatic calibration methods using an optimization technique. The manual method is time consuming and subjective, and does not include a procedure for seeking optimal values, nor does it provide associated calibration statistics for model analysis. In contrast, "automated" calibration is faster, less subjective, and can provide valuable statistics (Poeter and Hill, 1997).

All watershed models have parameters, which can be coefficients or exponents in the model equations. There are two main approaches to estimating these model parameters. The first approach (a priori approach) estimates model parameters by relying on theoretical or empirical relationships that relate such parameters to measurable characteristics of the watershed, such as soil and vegetation properties, watershed geomorphology, and topographical features. The second approach (model calibration) adjusts model parameter values, so that the model response closely matches the measured response of the watershed.

There are quite a few published contributions involving automatic calibration of distributed watershed-scale models. Users are referred to some of the recent contributions in the area. These include Refsgaard and Knudsen (1996), Refsgaard (1997), Senarath et al. (2000), Eckhardt and Arnold (2001), Muleta and Nicklow (2005), Boyle et al. (2001), and Ajami et al.(2004). The studies by Senarath et al. (2000), Ajami et al. (2004), and Eckhardt and Arnold

(2001) used a Shuffled Complex Evolution (SCE-UA) (Duan et al., 1992) search algorithm. Muleta and Nicklow (2005) used genetic algorithms (GAs) (Holland, 1975) for automatic calibration. Most of these methods used Monte Carlo approaches. In Monte Carlo method, a model is run multiple times, each time changing the values of the model inputs and parameters by randomly sampling from their associated population distributions. Output values from all simulations are collected, the distribution of values is plotted and analyzed and the impact of the input variation on the output variation is observed. The output distribution is compared with the distribution of the actual measurements.

There are also other studies that used multiple objective optimizations, which improve model calibration but increase the computational complexity. The classical multi-objective optimization involves transforming multiple objectives into a single function. Other studies implemented Pareto optimization to avoid the disadvantages of converting multi-objective functions into a single optimization problem (Yapo et al., 1998; Boyle et al., 2001; Vrugt et al., 2003; Khu and Madsen, 2005), by generating a set of solutions that provide equally good fit. Use of a single objective function is appropriate when objectives do not conflict. In this case a Pareto optimization will converge to a single set of parameter values. Multi-objective calibrations deal mainly with unrelated functions that do not conflict. Gupta et al. (1998) discussed the use of alternative error statistics as objective functions in the multi-objective calibration of a model to simulate one aspect (i.e., daily flow) of a watershed system. They used a general-purpose global multi-objective optimization algorithm (MOCOM-UA) and successfully estimated the Pareto solution space. They noted that the objectives should be unrelated functions that measure different aspects of the relationship between the data and the parameters of the system.

The case studies in Chapter 11.0 in this manual use UCODE nonlinear regression method to estimate parameters for the WARMF model of the Turkey Creek watershed given observations of daily stream flow measurements. Further information on hydrology and water quality calibration can be obtained from (Geza et al., 2008; Geza and McCray, 2008). UCODE (Poeter et al., 2005) performs inverse modeling, posed as a parameter-estimation problem, by calculating parameter values that minimize the sum of weighted-squared-residuals with respect to the parameter values using a modified Gauss-Newton method (used in this guide) or Trust-Region technique. The objective function is the sum of weighted-squared-residuals. Any application model or set of models can be used. An estimated parameter can be a quantity that appears in the input files of the application model(s), or that can be used in conjunction with user-defined functions to calculate a quantity that appears in the input files. An observation can be any quantity for which a simulated value can be produced, to which the regression will compare the observation. Simulated values can be extracted by UCODE from the application model output files. To use UCODE, the user must prepare a set of files that define parameters. Primary output is sensitivity of the observations to parameters and the estimated parameters that minimize the objective function. Other output includes statistics that describe how well the application models match the observations using the estimated parameters. The official version of UCODE can be obtained freely via U.S. Geological Survey water-resources analysis software or electronically from the internet (see Appendix C).

UCODE estimates parameter values that minimize the objective function. In its simplest form the objective function is the sum of weighted least-squared residuals between observed and simulated values given by:

$$O(\beta)=\sum_{i=1}^{N} w_i[y_i-\hat{y}_i(\beta)]^2 \qquad (9.3\text{-}1)$$

where: O(β) = the objective function,

β = the values of the parameters at which the function is evaluated,

y_i = the ith observation,

$\hat{y}_i$ = the ith simulated value equivalent to that observation and w_i = the weight associated with the ith observation (here the inverse of the measurement variance was used); and N is the total number of observations. The process requires sensitivities (discussed in Section 9.4), which can be calculated by perturbation if the process code does not provide them, which is generally the case.

Sensitivities are required because the objective function is calculated each time by perturbing the parameter value. Thus, $\hat{y}_i(\beta)$ in the above equation is calculated as:

$$\hat{y}_i(\beta)=\hat{y}_i(\beta_o)+\frac{\partial\hat{y}_i(\beta)}{\partial\beta}.(\beta-\beta_o) \qquad (9.3\text{-}2)$$

where: $\partial\hat{y}_i(\beta)/\partial\beta$ = a sensitivity matrix and parameter and β_o is the unperturbed parameter value.

The residuals (observation minus simulated value) are squared, weighted, and summed to produce the sum-of-squared-weighted residuals objective function. The objective function measures how well the model fits the observations. The goal of regression is to find parameter values that produce the smallest value of the objective function.

Sensitivities can be calculated by UCODE using perturbation methods. For each execution, one parameter value is increased or decreased slightly (perturbed) from its unperturbed value, while the other parameter values are not perturbed. The differences between the resulting perturbed simulated values and the unperturbed simulated values are used to calculate the sensitivities. Once the residuals and the sensitivities are calculated, they are used to perform one parameter-estimation iteration using the modified Gauss-Newton nonlinear regression parameter-estimation method described by Hill and Tiedeman (2007), which was modified from the method described by Hill (1998) and Cooley and Naff (1990).

The last step of each parameter-estimation iteration involves comparing two types of quantities against convergence criteria: 1) changes in the parameter values, where a unique criterion can be specified for each parameter; and 2) the change in the sum-of-squared-weighted residuals. If the changes are too large and the maximum number of parameter-estimation iterations has not been reached, the next parameter-estimation iteration is executed. If the changes are small enough, parameter estimation converges. If convergence is achieved because the changes in the parameter values or the changes in the objective function (depending which criteria is used) are small enough, the parameter values are more likely to be the optimal parameter values – that is, the values that produce the best possible match between the simulated and observed values, as measured using the weighted least-squares objective function.

Figure 9.2 shows a schematic of an inverse analysis procedure. The input parameters are initially estimated by conventional means (e.g., using available laboratory and field test results).

The next step is to run the numerical simulation model to generate simulated values. The simulated values are then compared to the field observations and a regression analysis is performed to minimize an objective function. The objective function quantifies the fit between computed results and observations and it is minimized by updating the set of input parameters needed to perform the numerical simulation. If the model fit is not "optimal," the procedure is repeated until the model is optimized.

9.4 Sensitivity Analysis

Sensitivity analysis is the process of determining the rate of change in model output with respect to changes in model inputs (parameters). Physically based distributed watershed models simulate a wide range of complex aspects. Watershed models are calibrated against measured data by adjusting the parameter values according to certain calibration criteria. These models represent the different physical phenomena occurring within the catchment. Proper calibration and validation require a huge amount of data (land use, soil, geological, topographic, meteorological, and hydrological) representing the various variables and so these models involve a large number of model parameters. It is not practical to use all parameters in calibration especially for distributed model with a large number of parameters. Thus, the number of parameters is reduced either by using prior information to set parameters values within a certain range using information about the watershed, by a parameterization technique or through sensitivity analysis.

It is necessary to identify model parameters that could be reasonably estimated based on field data alone, using prior information, to reduce the number of parameters to be determined via calibration. To this effort, Refsgaard (1997) suggested that the parameter values be assessed using field data as much as possible, fixing spatial patterns of parameters to simplify the calibration process.

Even after setting some of identifiable parameters to a fixed value, there could still be too many parameters to use in calibration. Another technique used in calibration is parameterization. Parameterization involves transferring model parameters values of a given spatial unit to other spatial units using coefficients based on the knowledge of the relationships between the spatial units. Natural basins mostly show a great variability and complexity with respect to soil, land cover, topographic characteristics etc. For instance, saturated hydraulic conductivity varies in nature over a wide range. It is impossible to include all this diversity in calibration and come up with different values for each layer or location and it may not be reasonable to use a single value throughout the basin. To avoid this, the parameter has to be changed simultaneously for all layers or all soils based on the changes made on one of the layers or soils if the properties of one layer or soil can be related to another. Thus, the most important task before calibration is to pose a tractable calibration problem by limiting the number of parameters for which values will be estimated.

Another alternative to reduce the dimensionality of the parameter space is to use sensitivity analysis on the model output. Through sensitivity analysis, the parameters that are nonessential in influencing the model response can be fixed to their prior values (Muleta and Nicklow, 2005; Christiaens and Feyen, 2002). Not knowing the sensitivity of parameters can also result in time being uselessly spent on non-sensitive ones. Focus on sensitive parameters can lead to a better understanding and better estimated values, and thus reduced uncertainty. Therefore, sensitivity analysis as an instrument for the assessment of the input parameters with respect to their impact on model output is useful for model calibration and for reduction of uncertainty.

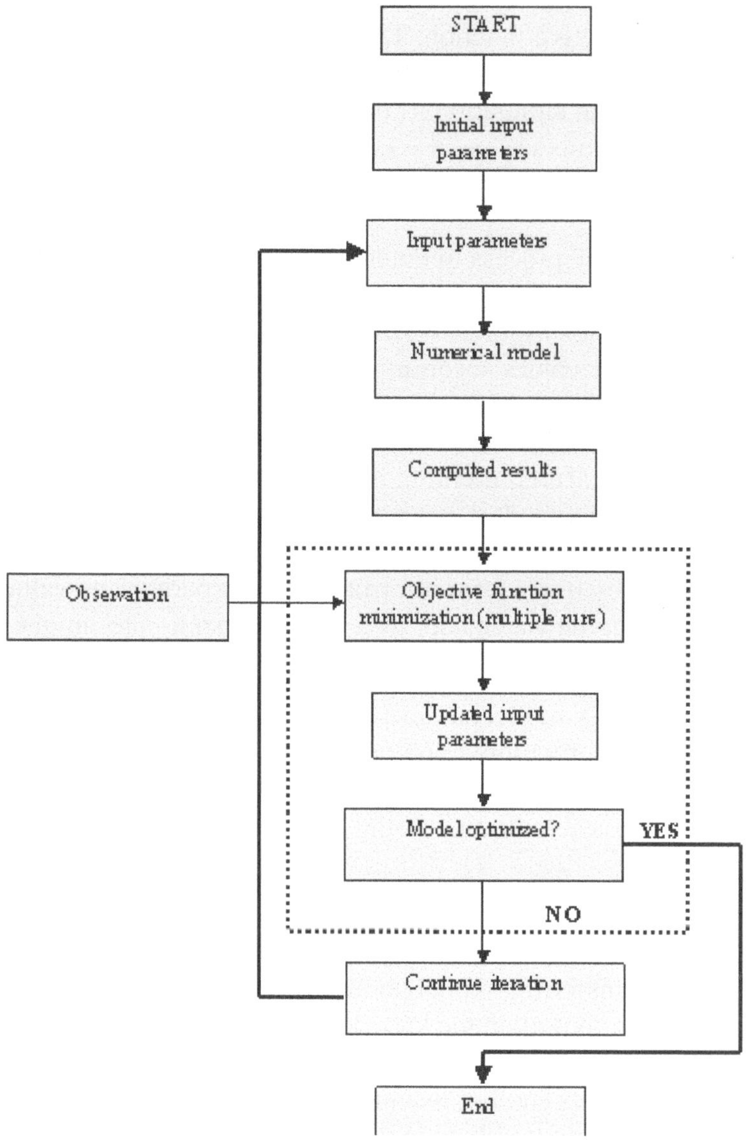

Figure 9-2. The Inverse Analysis Procedure.

A variety of methods are available for sensitivity analysis. Existing methods can be classified based on the way the parameters are treated (Saltelli et al., 2000). There are both local and global methods of analysis. The local methods estimate the local impact of a parameter value on the model output. Global techniques, on the other hand, analyze the whole parameter space at once. A local sensitivity analysis evaluates sensitivity at one point (e.g., a default value) in the parameter space. Sensitivities are usually defined by computing partial derivatives of the output functions with respect to the parameters. The sensitivity can be calculated for a small change of one parameter while the other input parameters are held constant.

The First Order Second Moment (FOSM) is a method of estimating the mean (first moment) and the variance (second moment) of model output through computation of the derivative of model output to model input at a single point (Yen et al., 1986). The Monte Carlo method evaluates the model for a large number of realizations of the problem. In this method, sampling is done from a possible range of the input parameter values followed by model evaluations for the sampled values. Global sampling methods are used to sample in a random or systematic way the entire range of possible parameter values and possible parameter sets. The simulation results are used to quantify the global parameter sensitivity or the uncertainty of parameters and outputs.

Essential to the Monte Carlo method is the sampling strategy (van Griensven, 2006). The Global sampling techniques include Monte Carlo sampling, Latin–Hypercube (LH) sampling, and variance-based methods. The Monte Carlo method provides approximate solutions to a variety of mathematical problems by performing statistical sampling experiments on a computer (Fishman, 1996). The samples are drawn from a specified distribution usually, a uniform distribution, which are then used to transform model parameters according to a predetermined transformation equation. An analysis of Monte Carlo simulations is conducted with statistical methods such as Kolmogorov-Smirnov (K-S) test (Stephens, 1970) to define whether a parameter is sensitive (Spear and Hornberger, 1980) or with the computation of regression and correlation based sensitivity measures (Saltelli et al., 1995). A great advantage of the method is the logical combination of calibration, identifiability analysis, sensitivity and uncertainty analysis within a single modeling framework (van der Perk and Bierkens, 1997).

The Latin-Hypercube simulation (McKay et al., 1979; Iman and Conover, 1980; McKay, 1988) is based on Monte Carlo simulation; however, the method uses a more structured sampling approach that allows efficient estimation of the output statistics. It subdivides the distribution of each parameter into n strata with a probability of occurrence equal to 1/n. For uniform distributions, the parameter range is subdivided into n equal intervals. Random values of the parameters are generated such that for each of the parameters, each interval is sampled only once. The Analysis of Variance (ANOVA) methods decomposes the variance into partial variances. A partial variance represents the first order effect of an input parameter on the output that corresponds to the variance when other inputs are kept constant. There are also higher order effects that combine the effect for 2 or more inputs. The partial effects can be calculated with special sampling schemes that are often computationally demanding (Saltelli et al., 2000).

UCODE (Poeter et al., 2005) used in the case studies presented in Chapter 11.0 performs local sensitivity analysis during model calibration to identify insensitive and correlated parameters. UCODE uses an alternative and computationally efficient first-order second-moment (FOSM) method, which directly propagates parameter uncertainty into predictive uncertainty. After the most relevant parameters are identified through sensitivity analysis, their values are estimated by calibration. UCODE can be used to calculate sensitivity of model output to input parameters so that insensitive and correlated parameters can be omitted from the parameter estimation to reduce the number of parameters by setting them to a constant reasonable value.

In UCODE, sensitivities can be calculated approximately using either a forward-, backward-, or central-difference approximation. For forward differences, each sensitivity (one for each observation with respect to each parameter) is calculated as:

$$\frac{\Delta y'}{\Delta b} = \frac{y'(b + \Delta b) - y'(b)}{(b + \Delta b) - (b)} \tag{9.4-1}$$

where: b = a vector (can be thought of as a list) of the estimated parameter values,

y'(b) = the value of the simulated value, y' , calculated using the parameter values in b,

Δb = a change in parameter value which sensitivities are being calculated (values for other are fixed),

y'(b + Δb) = the value of y' calculated using the parameter values at (b + Δb),

Δy' = the change in the simulated value caused by the parameter value change, Δb, and

Δb = a non-zero value which is called the perturbation for this parameter.

Although the regression uses sensitivities directly, scaled sensitivities are useful to the modeler. Dimensionless scaled sensitivities (DSS) indicate the importance of an observation to the estimation of a parameter as given by:

$$DSS_{ij} = \left(\frac{\partial \hat{y}}{\partial \beta} \right) \beta_j w_i^{1/2} \tag{9.4-2}$$

where $ß_j$ = the jth parameter (Hill and Tiedeman, 2007).

Composite scaled sensitivity (CSS) is the average of the DSS values associated with a parameter, and reflects the overall sensitivity of simulated equivalents to a parameter (Hill and Tiedeman, 2007).

$$CSS_j = \left(\frac{1}{N} \sum_{i=1}^{N} DSS_{ij}^2 \right)^{1/2} \tag{9.4-3}$$

Composite scaled sensitivity for a parameter can be used to evaluate whether the available observations are likely to contain enough information to allow estimation of a parameter. Parameters with larger CSS are more readily, and more precisely, estimated. The ratio of CSS of each parameter to the maximum CSS (γ_j) is given by:

$$\gamma_j = \frac{CSS_j}{\max(CSS)} \tag{9.4-4}$$

The ratio, γ_j, is useful for identifying parameters that can be estimated because it reflects the sensitivity of a parameter relative to the parameter with the highest sensitivity. γ_j varies from 0 to 1. The parameter with maximum sensitivity has a γ_j value of 1.0. γ_j less than about 0.01 are difficult to estimate and are associated with larger uncertainty parameters with γ_j less than about 0.01 are difficult to estimate and are associated with larger uncertainty (Hill and Tiedeman, 2007). A minimum value of 0.01 is used in this study to select parameters to be estimated through calibration.

9.5 Parameter Correlation and Uniqueness

Calibration of models of complex systems is commonly hampered by problems of parameter insensitivity and extreme correlation. Parameter correlation indicates the extent to which two random variables co-vary. Parameter correlation may result in non-unique estimated parameter values. Non-uniqueness implies that unique values cannot be estimated for the parameters. This suggests that there could be several combinations of parameter values that could give the same output. Parameter correlation often hampers calibration of watershed models. When parameters are correlated, the same result can be obtained with different combinations of values of the correlated parameters. In other words, if two parameters are correlated, increasing one and decreasing the other may achieve the same result.

A model that is calibrated with correlated parameters may not be sufficiently robust to provide good predictions for hydrologic conditions and time periods beyond those used for calibration. It is therefore essential to identify parameter values that can be uniquely estimated by determining the correlation between parameters. Parameter pairs with high correlation (e.g., absolute values of correlation coefficient greater than about 0.95) may not be uniquely estimated (Hill and Tiedeman, 2007). Uniqueness of the parameter estimates can be tested by repeating the parameter estimation with different initial/starting parameter values and assessing whether the same optimal parameter values are obtained. If the sum of squared weighted residuals achieved is similar and resulting parameter estimates differ from each other by values that are small relative to their calculated standard deviations, the optimization is likely to be unique. If this is not the case, the optimal parameter values are not unique.

Lack of uniqueness can be caused by a number of factors. If caused by local minima, it may be possible to examine the objective function value achieved by the different sets of parameter estimates and identify a global minimum as the set of estimated parameter values that is both reasonable and produces the smallest objective-function value. If non-uniqueness is caused by extreme parameter correlation, the objective-function value for each optimized set of parameters is likely to be similar and at least one pair of parameters will have a correlation coefficient very close to 1.0 or -1.0.

Correlation coefficients are reported by UCODE for all possible pairs of parameters. Parameter correlation coefficients (pcc) in UCODE are given by:

$$pcc = \frac{Cov\{b\}_{jk}}{[Var\{b\}_{jj}^{1/2}.Var\{b\}_{kk}^{1/2}]} \qquad (9.5\text{-}1)$$

where: $Cov\{b\}_{jk}$ = the covariance between two parameters and $Var\{b\}_{jj}$ and $Var\{b\}_{kk}$ are the variances of each of the parameters.

9.6 Model Validation

Model validation involves running a model using input parameters measured or determined during the calibration process. Validation procedures are similar to calibration procedures in that both try to match predicted to measured value. However, a dataset of measured watershed response selected for validation should be different than the one used for model calibration. The model parameters are not adjusted during validation but evaluated to see if they can reproduce good match between predicted and observed values.

These limits for adequate calibration are stricter than the rating for general model validation because model parameter values are optimized during calibration but not during model validation or application. Validation provides a test of whether the model was calibrated to a particular dataset or the system it is to represent. If predictions do not produce good results for the validation dataset, calibration and/or model assumptions may be revisited.

Generally, the objective of the validation step is to test performance of a calibrated model parameter set against an independent set of measured data. The measured validation and calibration data sets cover different time periods or conditions. Good validation results support the usefulness of the model to predict future conditions under alternative management scenarios and future climates. Since validation assesses the performance of a calibrated model parameter set against a set of independent measured data, it is typically more difficult to get a good validation performance in comparison to calibration. According to Refsgaard (1997), model validation is the process of demonstrating that a given site-specific model is capable of making "sufficiently accurate" simulations, although "sufficiently accurate" can vary based on project goals. According to the U.S. EPA (2002), the process used to accept, reject, or qualify model results should be established and documented before beginning model evaluation.

9.7 Model Evaluation

Model users would like to know how good the calibration results are and how to measure them. The model evaluation techniques discussed are used during model calibration and validation to see if simulation results are close enough to the observed data. There are both statistical and graphical model evaluation techniques that are applied to calibration and validation of models. The statistical methods may include standard regression, dimensionless and error index. Graphical methods can also be used to visually compare measured and observed values to get an overview of model performance. Few of the statistical methods include the slope and y-intercept, the coefficient of determination (R^2), Index of agreement (d), Nash-Sutcliffe efficiency (NSE), and error indices such as Mean Absolute Error (MAE), the Mean Squared Error (MSE), the Root Mean Squared Error (RMSE) and Normalized mean bias (NMB).

The slope and y-intercept of the best-fit regression line can indicate how well simulated data match measured data. The slope and y-intercept are commonly examined under the assumption that measured and simulated values are linearly related, which implies that all of the error variance is contained in simulated values and that measured data are error free (Willmott, 1981). The coefficient of determination (R^2) describes the degree of co-linearity between simulated and measured data. R^2 ranges from 0 to 1, with higher values indicating less error variance, and typically values greater than 0.5 are considered acceptable (Santhi et al., 2001, Van Liew et al., 2003). R^2 is oversensitive to high extreme values (outliers) and insensitive to additive and proportional differences between model predictions and measured data (Legates and McCabe, 1999).

The index of agreement (d) developed by Willmott (1981) is another measure of a standardized measure of the degree of model prediction error. It is calculated as:

$$d = 1.0 - \frac{\sum_{i=1}^{N} |Q_o - Q_s|}{\sum_{i=1}^{N} |Q_s - \overline{Q_o}| + |Q_o - \overline{Q_o}|} \qquad (9.7\text{-}1)$$

where; Q_o = measured stream flow,

Q_s = simulated flow and $\overline{Q_o}$ is measured mean stream. The index of agreement, d, varies between 0 and 1. A value of 1 indicates a perfect agreement between the measured and predicted values, and 0 indicates no agreement at all (Willmott, 1981).

The Nash-Sutcliffe Efficiency (NSE) (Nash and Sutcliffe, 1970) determines the model efficiency as a fraction of the measured stream flow variance that is reproduced by the model. NSE is given is calculated as:

$$NSE = 1 - \frac{\sum (Q_o - Q_s)^2}{\sum (Q_o - \overline{Q_o})^2} \qquad (9.7\text{-}2)$$

where: Q_o = measured stream flow,

Q_s = simulated flow and $\overline{Q_o}$ is measured mean stream flow. The closer the NSE value to 1.0 the better is the estimation of the stream flow by the model. NSE $\geq$ 0.75 is considered to be an excellent estimate, and NSE between 0.75 and 0.36, is regarded to be satisfactory (Motovilov et al., 1999).

Error indices including mean absolute error (MAE), mean square error (MSE), root mean square error (RMSE), the ratio of RMSE to standard deviation ratio (RSR), normalized mean bias (NMB) are frequently used model evaluation. RMSE, MAE, MSE, RSR and NMB values of 0 indicate a perfect fit. Singh et al. (2004) state that RMSE and MAE values less than half the standard deviation of the measured data may be considered low and that either is appropriate for model evaluation.

The mean absolute error (MAE) function is given by:

$$MAE = \frac{\sum_{i=1}^{n} |Q_s - Q_o|}{n} \qquad (9.7\text{-}3)$$

where; Q_s = simulated stream flow and Q_o = measured stream flow and n is the number of observations. As the name suggests, the mean absolute error is a weighted average of the absolute errors.

The mean squared error or MSE is one of many ways to quantify the amount by which an model predictions differs from the measured value of the quantity being estimated. MSE is called squared error loss. MSE measures the average of the square of the error where the error is the amount by which the model prediction differs from the measured quantity to be estimated.

Root Mean Square Error (RMSE) given by:

$$RMSE = \sqrt{\frac{\sum_{i}^{n} (Q_s - Q_o)^2}{n}} \qquad (9.7\text{-}4)$$

where: Q_s = simulated stream flow,

Q_o = measured stream flow and n = the number of observations. An RMSE value closer to zero indicates a better fit to observed values. RSR is the ratio of RMSE and standard deviation standardizes the RMSE.

Normalized mean bias (NMB) measures the average tendency of the simulated data to be larger or smaller than their observed counterparts (Gupta et al., 1999). Normalized mean bias (NMB) is used as a normalization to facilitate a range of values. This statistic averages the difference (model - observed) over the sum of observed values. NMB is a useful model performance indicator because it avoids over inflating the observed range of values, especially at low concentrations. Normalized mean bias is defined as:

$$NMB = \frac{\sum_{1}^{n}(Q_s - Q_o)}{\sum_{1}^{n}(Q_o)} * 100 \qquad (9.7\text{-}5)$$

where: Q_o = measured stream flow and Q_s is simulated flow.

Graphical techniques such as time series plots, percent exceedance probability curves, bar graphs and box plots, can also be used to compare seasonal variations and data distributions. Moriasi et al. (2007) reviewed recommended model evaluation techniques, both statistical and graphical. They reported ranges of values and corresponding performance ratings for the recommended statistic and gave recommendation based on the review results and project-specific considerations for stream flow and transport of sediment and nutrients. Based on this analysis, they recommend that three quantitative statistics – Nash-Sutcliffe efficiency (NSE), percent bias, and ratio of the root mean square error to the standard deviation of measured data (RSR), in addition to the graphical techniques – be used in model evaluation.

9.8 Prediction and Associated Uncertainty

Uncertainty in model outputs is the result of the lack of accuracy in model input parameters and the processes the model attempts to describe. Predictive uncertainties may arise due to simplification of the system, errors in the conceptual model, and inadequate quantity and quality of data. In other words, uncertainty can result from either natural variability, such as by unpredictable septic effluent concentration, evapotranspiration, or by both known and unknown errors in the input data, the model parameters, or the model itself. Imperfect knowledge of the values of parameters associated with these processes constitutes parameter error within a model. Natural variability includes both temporal variability and spatial variability, to which model input values may be subject. For example, the rainfall measured at a weather station within a particular model grid cell may be used as an input value for that cell, but the rainfall may actually vary at different points within that cell and may have a different mean. Errors in the model itself, also termed model structural error, may be the result of imperfect representation of processes in a model. Model structural error can be reduced through model improvement, whereas natural variability of input data is a property of the natural system, and is usually not reducible at the scale being used. Computed values differ from observed values, and the magnitude and frequency of these differences characterize the uncertainty of the model estimates.

Parameter values can never be completely accurate. There are also stochastic variables over which we don't have any control. Model simulations are usually performed with single-point estimate of input variables (e.g., a single denitrification rate for a catchment or soil layer) to give a single deterministic output value (e.g., nitrate concentration). Often, the variability in the data used in the models or the inaccurate representation of the real world by the models is overlooked. In reality, these point estimates belong to distribution that reflects the uncertainties in the data and models used.

Predictions are usually the ultimate purpose of a modeling study. Even models of relatively low predictive accuracy can be useful to decision makers if the predictive accuracy is appropriately quantified. It is, therefore, essential that modeling efforts be accompanied by an analysis of predictive uncertainty. Uncertainty analysis is used to examine how the lack of knowledge in model parameters, variables, and processes represented by equations propagates through the model structure as model output or forecasting error. Uncertainty of model predictions can be characterized using Monte Carlo simulation, inferential statistics, or regression. Monte Carlo simulation involves multiple runs of the simulation model each time changing the values of the model inputs and parameters by randomly sampling from their associated population distributions. After finishing the simulations, simulations that are deemed to sufficiently match the field observations are considered. It is assumed that gathering a large number of such model simulations, the distribution of predictions will reflect the prediction uncertainty.

The parameter distributions can be sampled randomly or preferentially and weighted (e.g., Latin hypercube sampling, LHS). Both random sampling and Latin hypercube sampling can be used in the generation of the mapping between analysis inputs and analysis results. In random sampling, a random value is sampled from each distribution specified for each uncertain model parameter. The random sampling approach assumes that the model input variables and model parameters are random. Distributions of these input variables and parameters are defined using expert judgment and by calibration exercises. Once defined, the actual values of these random variables and parameters are drawn from these distributions for each simulation time step. Many simulations are performed, each using values drawn from these probability distributions, to generate a distribution of output values. After simulating a considerable number of time steps, one has a set of equally likely outputs that define a probability distribution for each selected performance measure.

Unlike random sampling, Latin Hypercube Sampling (LHS) is a more structured sampling technique. In order to generate samples from input distributions, the range of probable values for each uncertain input parameter is divided into M segments of equal probability. If there are N parameters, there will be M^N possible random samples (combinations). The model is run for each random sample the outputs are analyzed statistically, for example for complementary cumulative distribution function (CCDF) that defines a probability of exceedance. Typically on the order of tens of thousands of model simulations are required to capture the predictive uncertainty.

One example that uses the Monte-Carlo approach is the Generalized Likelihood Uncertainty Estimation (GLUE) method (Beven and Binley, 1992; Beven and Freer, 2001) which is an extension to Spear and Hornberger's (1980) Generalized Sensitivity Analysis (GSA). In both GLUE and GSA methods, ensembles of model parameters are sampled from distributions, typically with independent uniform or normal distributions for each parameter. The model is then run with many such parameter sets, producing multiple sets of model output. These are used together to generate credible regions for model predictions. The generated model parameters can be grouped parameters that produced results consistent with the observed values and parameters that produced results that contradict the observations. These methods are based upon Monte-Carlo simulation. Parameter sets are sampled from a probability distribution, with most reported applications sampling from uniform distributions (Beven, 2001). Each parameter set is used to produce a model output; the acceptability of each model run is then assessed using a goodness-of-fit criterion. Muleta and Nicklow (2005) also used GLUE (Beven and Binley, 1992; Beven and Freer, 2001), which is based on Monte Carlo simulations. A major drawback of the GLUE methodology is the subjectivity of the likelihood level assignment that is used to group the parameter sets into the acceptable and non-acceptable (Muleta and Nicklow, 2005) categories. Lin and Radcliffe (2006) used the Parameter Estimator (PEST) software (Doherty, 2004) to estimate predictive uncertainty. The automatic model calibration and predictive uncertainty analysis based on a Monte Carlo approach were undertaken using a suite of utility codes written to support the use of PEST in the surface water modeling context. Monte Carlo based approaches for evaluating predictive uncertainty require many simulations to obtain a satisfactory description of the output distribution and are typically more time consuming than use of inferential statistics.

Inferential statistics is another approach to estimate uncertainty. If the model is fairly linear with respect to the parameters, predictive uncertainty can be estimated with only on the order of tens of model runs by using inferential statistics to calculate linear confidence intervals on predictions. Linear confidence intervals are symmetrical with respect to the predicted value. If the model is non-linear, then nonlinear regression can be used to calculate the nonlinear confidence intervals on the predictions, increasing the required number of model runs by twice the number of predictions for which uncertainty is evaluated.

The uncertainty analysis associated with UCODE uses inferential statistics and may include only the uncertainty associated with the estimated parameters, or additional input can be provided to account for uncertainty associated with parameters that were not estimated because of insensitivity, parameter correlation, or both. Including such parameters is important if their values are important to the values of the predictions. The linear uncertainty approach in UCODE can be used to calculate 95% linear confidence and prediction intervals on predictions. Confidence intervals represent the uncertainty in the simulated values that results from the uncertainty in the parameter values. Prediction intervals include the uncertainty in the parameter values as described for confidence intervals, and also include the effects of the measurement error that is likely to be incurred if the predicted quantity is to be measured. UCODE calculates both individual and simultaneous confidence or prediction intervals. Individual intervals apply when only one prediction is of concern. Simultaneous intervals apply when the joint uncertainty of more than one prediction is considered. Different types of simultaneous intervals are available including "Bonferroni," "Scheffé d = k," and "Scheffé d=NP" where k is the number of predictions and NP is the number of parameters. If k exceeds one and is less than NP, the smaller of the Bonferroni or Scheffé d = k simultaneous intervals is reported because both are conservative. If k is greater than NP, Scheffé d = NP simultaneous intervals are used (Poeter et al., 2005). Thus, for studies like this one Scheffé d = NP simultaneous confidence intervals is applicable. Linear confidence intervals are expressed as:

$$z'_\ell \pm [critical\ value] s_{z'_\ell} \qquad (9.8\text{-}1)$$

where z'_ℓ = the $z'_\ell{}^{th}$ simulated value for parameters that are estimated by regression

s'_{z_ℓ} (j) = the standard deviation of the ℓth predicted value is calculated as:

$$s'_{z_\ell}(j) = \left[\sum_{i=1}^{NP} \sum_{j=1}^{NP} \frac{\partial z'_\ell}{\partial b'_j} V(b) \frac{\partial z'_\ell}{\partial b'_i} \right]^{1/2} \qquad (9.8\text{-}2)$$

where: $\partial z'_\ell / \partial b'_j$ is the sensitivity of the ℓth prediction with respect to b_j, the jth parameter;

NP = the total number of defined parameters;

$V(b)$ = the symmetric square NP by NP parameter variance-covariance matrix for all parameters including those estimated in the regression and those for which uncertainty is specified by the modeler.

For parameters estimated by regression the variance-covariance matrix can be calculated as:

$$V(\underline{b}) = \frac{\text{sum of weighted squared residuals}}{ND - NP}\left(X^{T}\omega X\right)^{-1} \tag{9.8-3}$$

where: ND = the number of observed values,

X = the sensitivity matrix, or Jacobian,

NDxNP = matrix in which each element reflects the amount of change in one simulated equivalent to an observation with respect to change of one parameter value, and ω is an NDxND weight matrix equal to the inverse of the variance/covariance matrix of errors in the observations.

The critical value in equation (5) for scheffé d = NP simultaneous interval is given by:

$$\left[d \text{ x } F_{\alpha}(d,n)\right]^{1/2} = \left[NP \text{ x } F_{\alpha}(NP,n)\right]^{1/2} \tag{9.8-4}$$

where: α = the significance level (0.05), which results in 95 % confidence intervals,

n = the degrees of freedom and NP = the number of parameters for which sensitivities are used. Tables of the statistics from F distributions used to calculate the critical values are programmed into UCODE.

The Monte Carlo approach is conceptually simple, can consider any character of distribution of input parameters, and is not affected by non-linearity or discontinuities in the model. However, the method is computationally intensive. For example, the GLUE method (one of the most popular approaches) has frequently been criticized for its large computational demands. Kuczera and Parent (1998) note that GLUE "may require massive computing resources to characterize a highly dimensioned parameter space." Jia and Culver (2006) report generating 50,000 parameter sets to find 381 acceptable sets (just 0.8%). Use of a weighted sampling scheme can decrease the amount of computation, but still many more model simulations are needed than required when using inferential statistics.

The inferential statistics approach used in UCODE is computationally efficient and can be directly tied to certainty associated with observed values. The method is conceptually more complex, assuming the model is fairly linear with respect to the parameters (tests for linearity can be evaluated prior to assuming a more rigorous approach). Nonlinear regression can be used to calculate nonlinear confidence intervals on predictions. This increases the number of model simulations but continues to be substantially fewer than required for Monte Carlo approaches. Although parameter uncertainty can be directly incorporated into predictive uncertainty, scenario uncertainty (i.e., representation of future conditions) is still represented in a Monte Carlo style approach.

9.9 Monitoring Needs for Long-Term Use

Water quality monitoring involves recording data about various characteristics, and analyzing and interpreting the data. Monitoring helps ensure that a waterbody is suitable for its determined use. It can also be used for protective purposes to prevent degradation or to upgrade the class. Water quality monitoring can serve as a critical component of the restoration of the watershed and water quality. Monitoring serves multiple purposes: it estimates nutrients and sediment loads and it quantifies trends in concentrations that potentially reflect effectiveness of control practices. Monitoring also helps calibration of watershed models, verifying and improving its performance. Watershed modeling and monitoring activities should:

- be better integrated to reveal key uncertainties;
- provide focus for strategic research; and
- improve model accuracy and interpretation of the observations.

Simpler models may require fewer monitoring data relative to more complex models that have great data needs.

Monitoring is critical for model calibration and for identifying model inadequacies and lack of model fit. Thus, the modeling effort must initially rely heavily on monitoring data, and model and monitoring results should be frequently compared. It is critical to understand the limitations of using monitoring data to support modeling efforts. In conducting such comparisons it is important that the uncertainties associated with both the monitoring (i.e., how close monitoring data estimates are to "real" flows and loads) and the modeling (what are the model uncertainties as affected by input parameter uncertainties and model structure) be assessed and documented.

The predictive power of the model is only as accurate as our understanding and model representation of all the factors and processes that combine to determine transport rates. Because information validating the effectiveness of many of the practices is sparse, input parameters are highly uncertain and the model has large spatial lumping. It is unreasonable to attach a high degree of certainty to model predictions of future. A critical present need is to conduct an uncertainty assessment of the model and monitoring results to improve the general understanding of the degree of uncertainty associated with model predictions and the possible reasons for disagreement between model and monitoring results.

CHAPTER 10.0

RISK-BASED DECISION MAKING USING UNCERTAIN MODELS

10.1 Models Yield Uncertain Results

Many decision makers have relied on a single model output or "answer" to the question at hand, which enables a more straightforward decision. Some assume that model results can remove much of the uncertainty from a decision. However, making a decision based on model output without considering model uncertainty is akin to making a decision without considering all the relevant factors. Understanding the uncertainty in model outputs enables a more informed decision that can be tied to the risk associated with various potential outcomes.

10.1.1 Factors Leading to Model Uncertainty

Watershed model results are uncertain because:

(a) the mathematical model is only a simplified representation of the conceptual model of the watershed;

(b) the conceptual model of the watershed never represents all aspects of the actual watershed;

(c) several different conceptual models may be equally valid;

(d) data required for model input are not available;

(e) only a fraction of the data needed to parameterize and calibrate the model are available; and

(f) future conditions that may influence model predictions are uncertain.

Each of these topics is discussed in more detail throughout this chapter. Concepts in a) through c) are discussed in Section 10.A.4. Concepts in d) through f) are the primary focus of this chapter and each involve the same concept: evaluate model-input uncertainty and quantify the uncertainty in the model output as a result. Because of these factors, many quantitative hydrologists are strong proponents of avoiding calibration of watershed models in favor of a rigorous evaluation of model uncertainty. The uncertainties in model results are then used in risk-based decision making. This concept is described in Section 10.3.

10.1.2 Uncertainty in Calibrated and Non-Calibrated Models

Many professionals point out that calibrated models are fraught with uncertainty that is not typically evaluated in detail. Indeed, even highly rigorous calibration procedures require that many or most model-input parameters be fixed at "reasonable" values because calibration is possible only if a few model-input parameters are used. (Section 11.2.2 provides a good example of these concepts.) Rather, these professionals suggest that models are most useful when they are not calibrated, but instead run in a manner to rigorously evaluate uncertainty in model input and output. Other hydrologists are strong proponents of calibrating models, and this is the current standard best practice in the planning and regulatory arena surrounding OWS. Thus, most of this guide is dedicated to this approach. However, uncertainty analysis should be conducted for all models used in making decisions.

10.1.3 Uncertainty Analysis is a Best Practice

Both calibrated and uncalibrated models are frequently used to help in decision making. For both categories, a rigorous uncertainty analysis should be considered. Unfortunately, this aspect of model development is not usually included in current best practices. This chapter is devoted to this uncertainty analysis, for both calibrated and uncalibrated models. Classic and modified methods are provided. First, we have a brief discussion on the uncertainty due to alternate conceptual models.

10.1.4 Uncertainty Due to Alternate Possible Conceptual Models

Even when model calibration is conducted, the long-term predictive ability of a model is not likely to be robust if an incorrect conceptual model is used. However, there may be more than one equally correct conceptual model. Alternate conceptual models exist because we cannot completely characterize natural hydrologic systems. Different conceptual models are possible due to reasonable uncertainty in the structure of hydrogeologic units, boundary conditions, and model-input parameter sets (Poeter and Anderson, 2005).

Uncertainty due to alternate conceptual models, a) through c) in Section 10.1.1, can never be fully addressed, because a mathematical model can never completely address all aspects of even a single conceptual model. In addition, equally competent hydrologists and environmental scientists are likely to conceptualize equally valid – but different –representations of the same system.

It is important that stakeholders agree on the conceptual model(s) being used. The best-practice is to consider all conceptual models that are thought to be equally important. This practice is based on the lessons taught by legendary geologist T.C. Chamberlin, who believed that evaluation of "multiple working hypotheses" are the best way to make rapid advances in the study of both theoretical and applied problems (Chamberlin, 1890).

A rigorous approach to evaluate alternate conceptual models is provided by Poeter and Anderson (2006) (see Appendix C). The general procedure is to construct mathematical models for each alternate conceptual model. Both relatively simple and complex conceptual models can be considered, because simple models sometimes yield better results. The mathematical models can be ranked by objective means. For example, mathematical models that have better calibration statistics (see Chapter 9.0 on calibration) are ranked higher; models are discarded if calibrated parameter values are not consistent with observed hydrologic structures (e.g., if the calibrated hydrologic parameters for an observed sandy soil are representative of clay soils). (An example of the latter issue is given in the case study in Chapter 11.0, Section 2.2.) Many other criteria may also be used (Poeter and Anderson, 2006). Finally, the various mathematical model results are weighted according to the ranking of the model, and averaged. The average value might be considered the most likely answer (for example, the average daily nitrogen concentration), while the range of values produced by the highest ranked models might also be evaluated in the decision-making process.

10.1.5 Example of Model Uncertainty Analysis

This section provides an example of how model uncertainty analysis can be used in decision making. This example is geared primarily for the case when the model is not or cannot be calibrated because of lack of data. Later sections in this chapter provide more details on model uncertainty analysis for both calibrated and uncalibrated models. These later sections are intended primarily for those who will be conducting the modeling analysis.

A model uncertainty analysis will provide a quantitative measure of the uncertainty in model output, based on a specified uncertainty in model input. For example, aquifer hydraulic conductivity (K) is an important watershed hydrology parameter. However, even if a few measurements are available in the watershed, these few measurements cannot adequately represent the spatial variability across the watershed. If no measurements are available, then the uncertainty is likely to be several orders of magnitude! Another example is the denitrification rate, as described in Chapter 4.0, which enables removal of nitrate in the subsurface. A professional hydrologist or civil engineer with modeling expertise should be employed to evaluate model-input parameter uncertainty. This topic is addressed in a subsequent section of this chapter.

As an illustrative example, consider a model that is designed to calculate nitrogen concentrations in surface or groundwater at a particular location. The model is run repeatedly using appropriate distributions of model inputs that consider input-parameter uncertainty for all relevant input parameters. The collection for the output (N concentration) are called "ensemble." The ensemble can be arranged to provide a probability histogram (which can be constructed using a spreadsheet) similar to the one shown below in Figure 10-1.

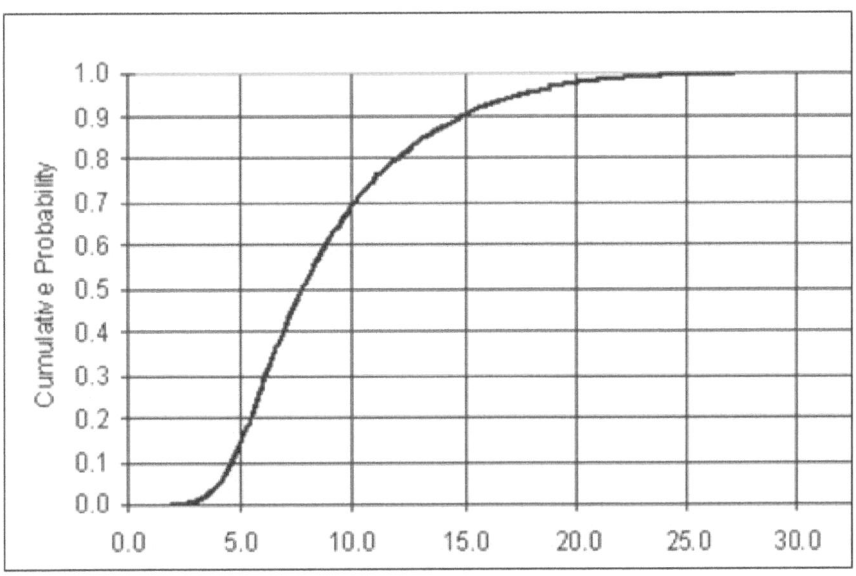

Figure 10-1. Cumulative Frequency Diagram Showing the Median Model Result, Range of Values, and the Probability of a Particular Model Outcome.

The figure depicts the range of model outcomes for nitrogen concentration and the theoretical probability that any particular outcome will occur. Of course, this assumes that a sufficient number of simulations are run (a topic discussed in more detail in Section 10.4). For example, the 50th percentile value of all model predictions is read from the y axis. This is called the median case, and is considered the most likely model prediction. The model prediction is 7.5 mg/L for this case, which is below the maximum contaminant level (MCL) for drinking water, but significantly above background. However, the 70th percentile concentration is 10 mg/L (the MCL). One way to interpret this result is to conclude that there is a 30% probability that the MCL will be reached or exceeded, based on the model simulations. This information can lead to an informed, risk-based decision, as described in the next section.

10.2 Relationships Between Model Results and Willingness to Accept Risk

Models results are frequently employed by planners and regulators to aid decisions regarding the potential impacts of OWS, or to try to determine whether OWS has been partially responsible for current impacts. Examples of relevant model output results include total maximum daily load (TMDL) of phosphorus in a stream segment, nitrogen concentrations in groundwater at a certain location, and increase in nutrient concentrations in streams, lakes, or groundwater. Typical decisions include whether or not to allow a development to proceed using OWS for wastewater treatment, whether to convert OWS to sewers, or whether to require advanced treatment units for existing and proposed OWS.

Any decision based on model-simulation results carries some risk that the correct model result was not obtained, and therefore that the resulting decision was inappropriate. This risk can be quantified and managed if a proper model uncertainty analysis is conducted.

Quantifying model uncertainty leads to a more thorough, informed, and defendable risk-based decision-making process.

In the example above, the stakeholders are presented with the result that the most likely outcome based on model simulations is that an MCL will not be exceeded. However, there is a significant possibility that an MCL will be exceeded (30% probability in the above example). This information complicates the decision, but leads to a better decision.

Some stakeholders will undoubtedly be uncomfortable with accepting the risk of violating a water quality standard based on this information. Others may argue that this risk is acceptable in a practical world. The decision makers must determine whether they are willing to accept the risk that one of the cases in the upper 30 % of simulations is most representative of the true conditions, rather than the assumed "most likely" case.

While this approach appears to make the decision-making process more complicated, it has considerable advantages. For example, this type of information can result in a stakeholder agreement to proceed with a decision that assumes an MCL will not be exceeded, provided that water quality monitoring is conducted, and that contingency plans are made ready for rapid implementation. In summary, the evaluation of the uncertainty can help stake-holders implement a decision more rigorously and effectively.

On the other hand, if the result had been different, and the 10 mg/L value instead occurred at the 95th percentile value, then perhaps a strong majority would be more willing to accept the apparently small risk of violating a water quality standard. For this case, the final decision might be the same, but the associated requirements might be less demanding.

Section 4.10 provides another example of how model uncertainty analysis aided in decision making using a simple approach. This type of approach is also discussed in more in

Section 10.E.3. All methods strive to deliver the same product: an understanding of how uncertainty in the model or the model input parameters results in uncertainty in the model output variable of interest upon which a decision is based. There is no strict formula for how this process is undertaken, and the process is likely to differ depending on the scenario being evaluated, he stakeholders, and the decision that is under consideration.

10.3 Uncertainty in Calibrated Models

The goal of every hydrologic modeler is to rigorously calibrate his or her model to watershed data. Then, the model is presumed to have the ability to predict into the future. However, it is not always possible to calibrate models in a manner that is acceptable to all parties. For example, calibration data (stream flow, groundwater hydraulic head values, pollutant concentrations) may not be available in a watershed, or may not be available in sufficient quantity to enable ready acceptance of a watershed model by a majority of stakeholders. In addition, distributed watershed models generally have too many input parameters compared to the available data to uniquely calibrate the model (more detailed discussions on this topic is found in the chapter on calibration, Sections 9.3 through 9.5, and in the case study in Section 11.1.6., and in Poeter and Hill (1997). Even for calibrated models, there is a range of values for each input parameter that will provide an equally valid calibration. All these factors lead to uncertainty in the predicted model output parameter for calibrated models. If this uncertainty is quantified, then it can be used in decision making similar as described above.

For calibrated models, we wish to quantify the model's prediction uncertainty. In this guide, the researchers used a computer program called UCODE (Poeter and Hill, 1997) to implement calculations to quantify prediction uncertainty. UCODE is robust because it can be used with any model that provides text or ASCII output, and that accepts text input (most models). It can also perform calculation related to other important aspects of model performance evaluation, including sensitivity analysis, correlation of input parameters, and automated calibration (all discussed in Chapter 9.0, and in the case studies in Chapter 11.0). However, another software program, or manual calculations, can also be used to calculate prediction uncertainty.

UCODE uses the confidence interval (CI) approach to compute prediction uncertainty. A simple definition of a confidence interval is the numerical interval around a predicted parameter (e.g., nitrate concentration or daily load of phosphorus in a stream) that bounds the actual value of that parameter, within a specified degree of probability. The width of the CI is a measure of the uncertainty about the position of the true value of the estimated parameter. The degree of confidence is linked with the width of the CI.

In traditional normal statistics, the CI is the simulated value plus or minus some multiple of the standard deviation from the most likely value. In normal statistics, this is the population mean. In model-performance evaluation, the most-likely value is the simulated value obtained via calibration. The larger the range about the predicted value, the higher the confidence interval because a larger range provides more confidence (a higher probability) that the true value actually exists within the range. For example, a 68% CI is associated with a range of +/- 1 standard deviation. To achieve a 95% confidence interval, the associated range is 3 standard deviations about the most-likely value.

For model prediction-uncertainty analysis, the interval around the predicted value depends on the probability desired (usually 90% or 95%), the number of parameters and observations used in the prediction, and a standard deviation of the prediction. Once the confidence interval is calculated, the uncertainty can be determined.

10.3.1 Calculating Prediction Uncertainty for Calibrated Models

This guide utilizes the automated software UCODE to calculate prediction uncertainty. Other software can also be used, or the calculations discussed in this chapter can be conducted manually, although this would be very time consuming. More details on the calculations used in UCODE can be found in Poeter and Hill (1997) and Poeter et al. (2005).

UCODE can calculate linear or nonlinear confidence and predictions intervals, which quantify the uncertainty of model simulated values depending on whether the model is linear or non-linear. The linear uncertainty can be used to calculate 95% linear confidence and prediction intervals on predictions. Confidence intervals represent the uncertainty in the simulated values that results from the uncertainty in the parameter values. Prediction intervals include the uncertainty in the parameter values as described for confidence intervals, and also include the effects of the measurement error associated with the predicted quantity (e.g., stream flow or nutrient concentrations). Linear confidence intervals on predictions are expressed as:

$$z'_\ell \pm [\mathit{critical\ value}]\, s_{z'_\ell} \qquad (10.3.1\text{-}1)$$

where: z'_ℓ = the z'_ℓ th simulated value for parameters that are estimated by regression and s'_{z_ℓ} (j) = the standard deviation of the ℓ th predicted value is calculated as:

$$s'_{z_\ell}(j) - \left[\sum_{i=1}^{NP} \sum_{j=1}^{NP} \frac{\partial z'_\ell}{\partial b'_j} V(b) \frac{\partial z'_\ell}{\partial b'_i} \right]^{1/2} \qquad (10.3.1\text{-}2)$$

where: $\partial z'_\ell / \partial b'_j$ = the sensitivity of the ℓ th prediction with respect to b_j, the j th parameter;

NP = the total number of defined parameters for which sensitivity is calculated and $V(b)$ = the variance-covariance matrix (symmetric with size NP by NP) for all parameters including those estimated in the regression and those not estimated but for which uncertainty is specified by the modeler. The critical value in equation (1) is given by:

critical value = [NP x Fa (P, n)]1/2 (10.3.1-3)

where: F = the "F distribution" from traditional statistical tables,

F_α = the F-value for a specified significance level α,

n = the degrees of freedom (NP and n are the values used to enter the F-distribution table, which is programmed into UCODE),

($1-\alpha$) = the confidence interval (e.g., $\alpha = 0.05$ results in a 95 % confidence interval and $\alpha = 0.1$ results in 90% confidence intervals, etc). The value for n (degrees of freedom) is calculated based on the number of parameters that are being estimated compared to the number of observations available for calibration. Poeter and Hill (1997) describe how to implement this technique in significant detail. In addition, the case study presented in Section 11.2.2 demonstrates an application of this technique.

10.3.2 Reducing Prediction Uncertainty

It is important to realize that, for calibrated models, the input-parameter uncertainty associated with factors such as spatial variability and measurement error are not considered explicitly because input parameter values are "calibrated" or fixed at values estimated outside of calibration (see Chapter 9.0). Rather, the sensitivity of the model output to variations in model-input parameter values is used to generate the range of uncertainty in model output. This approach addresses the uncertainty of the calibration procedure. The resultant statistics (e.g., confidence intervals) is termed inferential statistics (Poeter and Hill, 1997).

The uncertainty in the model prediction is based primarily on how sensitive the model is to specific input parameters (which varies depending on the hydrologic system and the specific mathematical being used), on the number of parameters being estimated compared to the number of observations, and the specified accuracy of the data used for calibration. The last factor cannot be improved significantly in most cases. The uncertainty associated with a particular prediction can generally be improved by collecting more observation data (both in the time and space domains) to be used in calibration, getting more detailed and reliable measurements for input parameters so the number of estimated parameters can be decreased, and sometimes by refining the mathematical model to better match the conceptual model of the hydrologic system.

10.3.3 Examples for Prediction Confidence Intervals

Prediction confidence intervals can be highly useful to decision makers because a statistical measure can be associated with the prediction. For example, consider the case where the model result is the average daily load for a certain time period (kg/day) at a specific location in a stream, with application to a TMDL. The model might be used to estimate how new OWS in a proposed development might impact the TMDL on a stream reach. Stakeholders might agree in advance upon a confidence level required for the model predictions (e.g., 90% CI). Then, the 90% confidence interval could be computed and applied to the model output value to assist in evaluating the model results. For example, suppose that the model simulates a 500 mg/day nutrient load, with a 90% confidence interval of +/- 100 mg/day. Then, decision makers can decide if a predicted load of 400-600 mg/day is acceptable for the specific management scenario. If the 90% confidence interval is too large, say +/- 450 mg/day, yielding a range of 50-950 mg/day for the model result, then additional data could be collected, more input parameters could be measured and fewer estimated, or the mathematical model could be refined, to try to improve the prediction.

Often, many predictions are associated with a particular management scenario (e.g., daily averages of pollutant load or concentrations). A confidence interval may be calculated for each of these predictions. An efficient way to present this information is in a cumulative frequency diagram, as shown in Figure 10-2 below.

The figure can be analyzed as follows: 50% of the predicted values (read from y axis) have a calculated prediction error of less than about 15% (corresponding value from x-axis), 80% of the simulated predictions have a calculated error of less than 23%, etc. This helps a user to evaluate the overall prediction uncertainty associated with a watershed model. If higher certainty in model prediction is desired, then additional data collection is likely necessary.

10.3.4 Advantages of the Prediction Uncertainty Approach

Calculation of uncertainty using this approach requires less computational effort and is more realistic when based on inferential statistics rather than Monte Carlo simulation approach (discussed in the next section). It is preferable that the statistics that quantify uncertainty be calculated using parameter values that provide the best fit to observations of the system, but the method could be applied to parameter distributions as defined for Monte Carlo simulation without calibration.

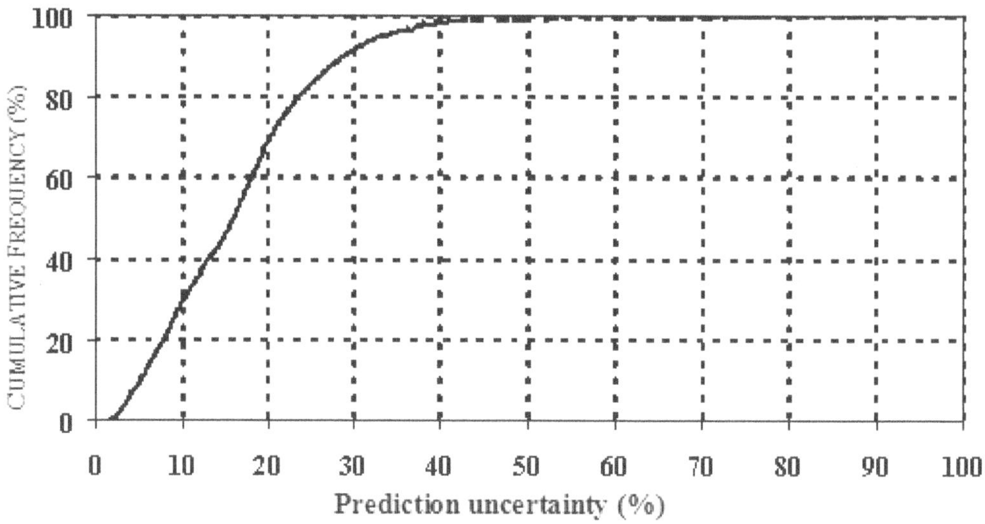

Figure 10-2. Prediction Uncertainty versus % of Predictions with Uncertainties Less than CFD.

10.4 Uncalibrated Models and Stochastic Approaches

10.4.1 When are Uncalibrated Models Appropriate?

Uncalibrated models are frequently used for decision making. This guide focuses primarily on model calibration because this is the state of the practice in the consulting, regulatory, and legal arenas. However, use of uncalibrated models is often useful, and sometimes necessary. In addition, as stated previously, some hydrologic modelers favor uncertainty analysis such as Monte Carlo simulations or other stochastic methods in lieu of model calibration, so this type of model analysis is becoming more common. We discuss Monte Carlo simulations, as well as a simplified approach to uncertainty analysis.

Data for calibration is not always available, especially when the model is being used to evaluate a potential scenario such as future development (recall Section 4.10). In addition, simple mass-balance spreadsheet models or GIS calculations that do not account for most physical, chemical, and biological processes can be used for screening (these models usually yield a conservative result). These models are designed to determine if the next step to more complex modeling is warranted (recall Sections 4.10 and 5.0).

When sufficient data for rigorous calibration are not available (this often cannot be determined until calibration is attempted), then an uncertainty analysis such as Monte Carlo simulations can be highly useful. This approach has been adopted for several of the hydrologic models used to evaluate suitability of Yucca Mountain as a nuclear waste repository. A simpler but less rigorous analysis was offered in Section 4.10 as an alternative to Monte Carlo Simulations. In either case, the first step is to determine the uncertainty in model input parameters.

10.4.2 Stochastic Models

Stochastic methods rely in using probabilistic ranges for model-input parameters, and conducting a large number of model runs to obtain a probabilistic range of model outputs. Monte Carlo simulations are a common stochastic method. This approach is touted by many hydrologic modelers because the spatial and temporal variability in many model input parameters or forcing functions used as input to watershed models are impossible to characterize. Extreme variability of natural hydrologic systems exists due to spatial variations in hydrologic and pollutant-transport processes, geological materials, and biological and chemical materials in watersheds. Temporal variations in hydrologic-flow and pollutant-transport compound this variability. Thus, watershed modeling is fraught with uncertainties. Stochastic methods are an attempt to rigorously address this uncertainty. Yet, watershed modeling has primarily relied on deterministic approaches, which is the primary focus of this guide.

However, a large volume of research has been compiled in the past 10 years that shows some clear advantages to stochastic methods (as opposed to deterministic) methods. An additional benefit is that stochastic methods are inherently suited to risk assessment similar to that discussed in Sections 10.3 and 10.4 above. However, stochastic approaches are not widely used in real-world applications, likely due to a lack of background, training, and tools.

In addition, even when a watershed model is calibrated, the inputs required to simulate future scenarios (climate, actual growth, etc.) are not available. Because simulation of future scenarios is often an important component of planning and decision making, it is important to be able to quantify the uncertainty in future hydrologic inputs. There are relatively simple ways to incorporate this uncertainty, so the researchers provided examples in the future-scenario simulations in Chapter 11.0. However, a more rigorous stochastic analysis is appropriate in many cases (e.g., variability in climate for long term predictions).

It is beyond the scope of this guide to discuss stochastic methods in detail. Indeed, most university programs in hydrology do not cover this topic. The following books are recommended as excellent text books to better understand this topic. The first one is a handbook with only a single chapter on stochastic methods, so it is a good place to start.

- Handbook of Hydrology, 1992: D.R. Maidment, Ed. McGraw-Hill, ISBN 0070397325.
- Handbook of Applied Hydrology, Ven Te Chow, McGraw-Hill, 1988, 712p, ISBN 00701.
- Stochastic methods in hydrology, 1996: rain, landforms, and floods: CIMAT, Guanajuato, Mexico, editors, O.E. ISBN 9810233671.
- Stochastic Methods in Subsurface Contaminant Hydrology, ASCE Press, 2002, Rao S. Govindaraju (Editor) ISBN 0784405328.

However, the basic concepts of stochastic Monte Carlo simulations are not that complex. In the remainder of this section, we provide a summary of the Monte Carlo procedure. The next section includes some examples of determining input-parameter uncertainty, which is the first step, and probably the most important and difficult step, to conducting Monte Carlo simulations. Understanding input-parameter uncertainty can also be useful for conducting a simplified uncertainty analysis as described in Section 4.10.

The next step in Monte Carlo simulations is to randomly sample from all the input-parameter distributions based on the probability of a particular value occurring, and run the model forward using the sampled set of input parameters. The model must be run many times (i.e., often thousands of times) for this procedure to be theoretically valid. The theory and procedures associated with this sampling process is rather involved and computationally time intensive. The textbooks listed above provide an informative treatment of this topic. One of the disadvantages of the Monte Carlo approach is that it is computationally intensive.

For the method to be statistically valid, all possible input-parameter combinations must be simulated or adequately represented. For watershed models such as the ones used in this guide, there can be more than 60 input parameters for each computational element (a sub-catchment). So, tens of thousands of model simulations can be required. The Generalized Likelihood Uncertainty Estimation (GLUE) method has frequently been criticized for its large computational demands. Kuczera and Parent (1998) note that GLUE "may require massive computing resources to characterize a highly dimensioned parameter space." Jia and Culver (2006) report generating 50,000 parameter sets to find 381 acceptable sets (just 0.8%).

Given that a typical watershed model that considered hydrology and pollutant transport may require several to dozens of hours to run on a typical PC, a super-computing facility would be required to properly run Monte Carlo simulations. Much research has been devoted to this topic in recent years,. For example, the Latin Hypercube Sampling (LHS) methodology has been used to obtain relatively accurate results without using all possible combinations of input parameters. While LHS can greatly reduce computation time, it is still significant, and thousands of simulations can be required to produce a theoretically valid ensemble of model outputs.

The next section provides examples of statistical distributions of input parameters. Monte Carlo simulation output, and use of the output in decision making, is similar to that described in Section 10.3 above.

10.4.3 Distributions for Model Input Parameters

10.4.3.1 Quantifying Uncertainty in Model Input Parameters

In a complex watershed model, not all model-input parameters are equally important, and therefore not all input-parameters need to be explicitly considered in an uncertainty analysis. Details on this topic are discussed in Section 9.4 on sensitivity analysis, and especially in the case studies presented in Section 11.2.2.

For each model input parameter, the uncertainty distribution for that parameter must first be determined. Three aspects must be considered: the most-likely or median value for the parameter; the variability in that parameter quantified by statistical parameters such as standard deviation or quartiles; and the shape of the distribution of the unknown parameters (e.g., normal distribution, log-normal distribution, and many others). Determining an accurate input-parameter distribution is a key to uncertainty analysis, but is not an easy task. If data are available, then the distribution can be determined directly. Examples include numerous measurements of hydrologic conductivity or phosphorus sorption capacity in a watershed, or data mined from the literature for general or specific watershed conditions.

Examples of the latter approach above are provided by McCray et al. (2005). More details on how to obtain these papers are listed in Appendix C. Figure 10-3 summarizes the results of an extensive literature review and data mining exercise by McCray et al. (2005).

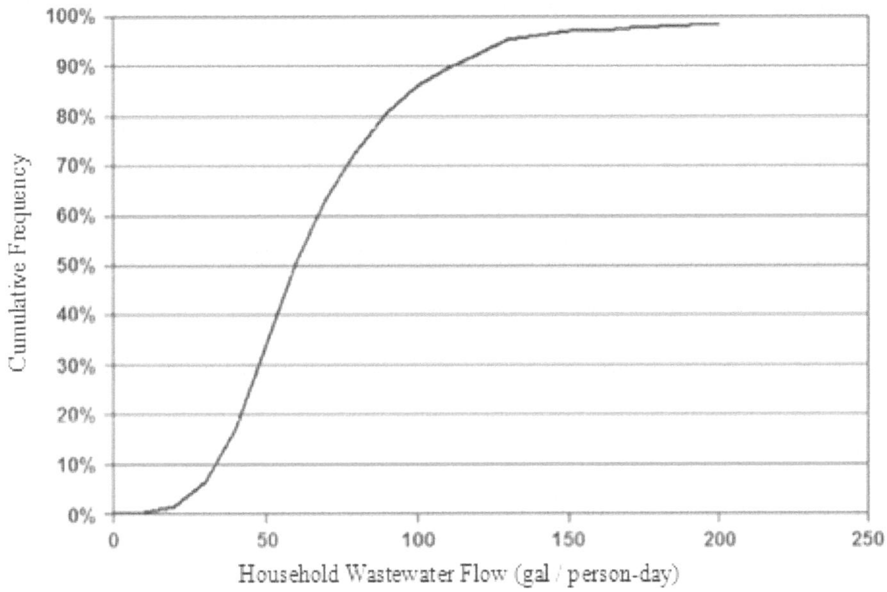

Figure 10-3. Cumulative Frequency Diagram for Household Wastewater Flow.
From McCray et al., 2005. Reprinted from *Ground Water* with permission of the National Ground Water Association, ©2005.

The figure demonstrates the variability of reported household wastewater flow rates. For example, the median flow rate of all households is about 60 gallons per person per day. For modeling evaluations, the median flow rate could be interpreted as the most likely flow rate for any particular household. The 80th percentile flow rate is about 95 gal/person-day (only 20% of the reported rates were higher than this value).

Figure 10-4 demonstrates a CFD for the first-order denitrification rate, one of several important model-input parameters typically needed for watershed modeling assessments of nitrogen loading to streams or groundwater. The denitrification rate parameter is responsible for conversion of nitrate to nitrogen gas in a model. The data for this figure is also taken from McCray et al. (2005), but is re-plotted on a semi-log scale because of the high variability in denitrification rates. The cumulative frequency is on the horizontal axis on this plot. The data compiled for this plot does not discriminate between soil types, although recent research suggests that denitrification rates likely depends on soil types. Tucholke et al. (2007) determined that denitrification rates were higher in finer-grained soils, and when water contents were higher (the soil is wetter). Details on how to obtain the paper are listed in Appendix C.

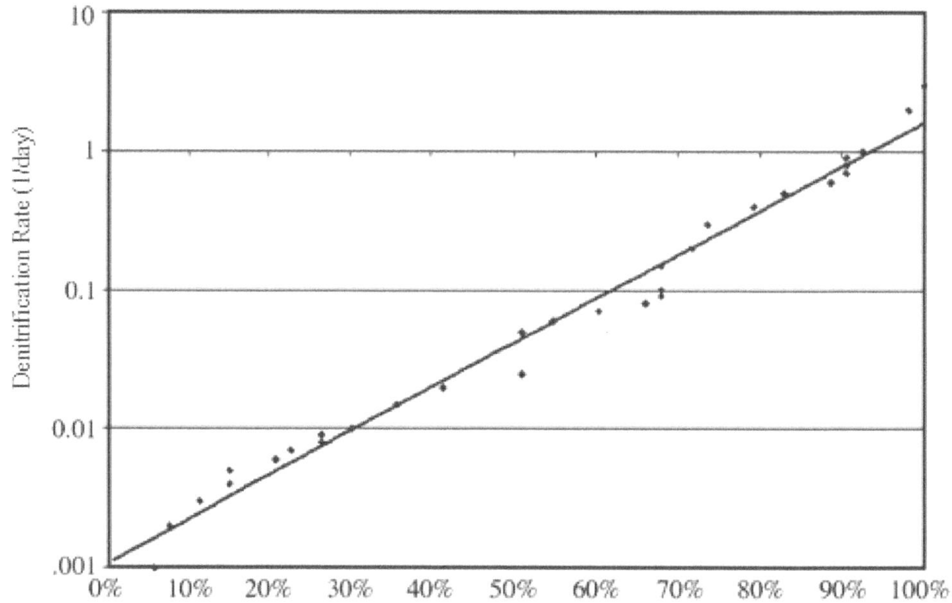

Figure 10-4. Cumulative Frequency Diagram of Denitrification Rates.
From Heatwole and McCray, 2007. Reprinted with permission from Elsevier.

Figure 10-4 shows that the median reported denitrification rate is about 0.05 day^{-1}. 90% of reported rates are less than about 1 day^{-1} and 90% of reported rates are greater than about 0.002 day^{-1}. This plot demonstrates that this parameter is highly uncertain. Moreover, most watershed models are very sensitive to either the denitrification rate, the nitrification rate, or both (see the case studies in Section 11.2.2). Denitrification rates are also very difficult to measure on the watershed scale, or even at the site scale, for that matter. Thus, choosing any one parameter for model input is not appropriate.

Table 10-1 lists the different model-input parameters that were evaluated by McCray et al. (2005), and states whether a CFD or other statistics were use to summarize the variation in that parameters.

10.4.3.2 Using Input Parameter Distributions in Monte Carlo Simulations

In the above examples, Monte Carlo simulations would randomly sample from these distributions (as well as distributions for other input parameters) to obtain a set of input parameters for a single run. This sampling procedure is performed numerous times to obtain an ensemble of equally likely model outputs (two examples of sampling procedures were briefly discussed earlier in this chapter). Then, the outputs are compiled in a similar fashion as in the CFD diagrams above to determine which model result is most likely (e.g., the median case), and what value are statistically high (e.g., the 90th percentile model-output value), or statistically low (e.g., the 10th percentile model-output value).

Table 10-1. Summary of Model Input Data from McCray et al. (2005).

Phosphorus concentrations in STE	CFD, range, median, n [1]
Phosphorus sorption coefficient	CFD, range, median, n
Maximum P sorption capacities by soil type	Range, median, n
Ammonium-N concentrations in STE	CFD, range, median, n
Nitrate-N concentrations in STE	Range, median, n
First-order denitrification rate	CFD, range, median, n
First-order nitrification rate	CFD, range, median, n
Zero-order nitrification rate [2]	Range, median, n
Organic N	Range, median, n
Total N	Range, median, n
Wastewater flow rate	CFD, n, range, median, n

1. CFD = cumulative frequency diagram, n = number of data

2. Tucholke et al. (2007) & Tucholke (2007) provide detailed information about zero-order denitrification rates for various soil types and conditions.

10.4.3.3 Using Input Parameter Distributions in Simplified Model Analysis

Another way to use input-parameter distributions in modeling and decision making is to choose some end member cases. For example, one could determine the most sensitive and most uncertain model-input parameters, and vary the range of those input parameters in a subjective fashion to better understand and quantify the likely variability in model output. This procedure was followed by McCray and Heatwole (2007) and is summarized in Section 4.10. (Details on how to obtain the paper are listed in Appendix C). For example, one might run 10 different model simulations where the model-input parameters are carefully chosen from the CFDs. Stakeholders generally should be involved in what CFD values are used. A "most-likely" case could be run using median values.

Similarly, stakeholders could agree on model input parameters that represent a certain probability of occurring based on the CFDs, and run the model using those values to determine if a water-quality violation, or other undesired impact, is predicted by the model. For example, one might want to be conservative with respect to water quality and assume a 60% CFD value for wastewater flow in a watershed, a 60% CFD value for nitrogen concentrations in STE, and a 25% CFD value for denitrification rate. This combination, while somewhat arbitrary, would yield higher concentrations of nitrogen compared to using median values for each parameter. Of course, these CFD values would need to be agreed upon by the stakeholders prior to implementation of a model. Most regulatory agencies do not account for any denitrification in model evaluations of nitrogen impacts from OWS, but rely on dilution and mixing to attenuate N concentrations.

Alternately, one could determine what CFD values result in predicting a violation of a water quality standard or adverse impact, and then stakeholders would decide whether they were comfortable with accepting those CFD values. This was the approach taken by stakeholders in the case study presented by McCray and Heatwole (2007) (see Appendix C).

10.4.3.4 Input Parameter Distributions for Lumped- and Distributed-Parameter Models

Lumped-parameter watershed models use one set of equations and one set of input parameters for an entire watershed. Nearly all analytical mass-balance models (typically implemented in spreadsheets) are lumped-parameter models.

Distributed models require input parameters that vary spatially (i.e., parameters are different in each computational element or model cell) based on variable hydrologic conditions throughout the watershed. Numerical models such as SWAT, WARMF, and MODFLOW are considered distributed models. However, most numerical models have some lumped parameters.

Most soil and land-surface parameters are distributed, while many climate parameters are lumped. In addition, because there are numerous input parameters for a watershed model, many parameters that are distributed are actually assigned a single value for all computational elements in the watershed. Thus, in effect, this parameter becomes a lumped parameter.

For both types of models (distributed and lumped parameter), one must determine a median value for each input parameter that is representative of the entire watershed, the shape of the distribution, and a characteristic measure of the spread or range of values within this distribution. For a normal distribution, an appropriate measure would be the standard deviation.

For lumped-parameter models, the single most appropriate parameter value must be representative of the entire watershed, so the average or median value from the distribution is appropriate. This value is most appropriately obtained from evaluating measurements of each parameter over the entire watershed. Ideally, at least 15 such measurements are available. Another approach is to take values from the literature, such as was done by McCray et al. (2005). However, for parameters related to the soil type or hydrogeologic materials, it is always better to have information that is specific to that soil type or geologic material.

The distribution shape for input parameters in a lumped-parameter model is a normal distribution. This is according to the Central Limit Theorem in traditional statistics, which states that potential means from a population are normally distributed even though the distribution of the population itself may have a different distribution. There are some statistical methods that can estimate the standard deviation for a normal distribution of means based on the statistics of the population. The reader is referred to a geostatistics textbook for further details (Isaaks and Srivastava, 1989).

For distributed-parameter models, populating models with input-parameters using a stochastic method is more complex. It is usually not possible to obtain a distribution of parameters for each computation element (sub-catchment for a watershed model or model cell for most groundwater models) in the watershed. Thus, a distribution for each input parameter is obtained that represents the entire watershed (as described above). The distribution is sampled randomly and different parameters values are assigned to each computational element for each model run. For this case, it is very important to know the distribution of the population for each model-input parameter because the values assigned to each computational element are intended to represent the spatial variability of that parameter in the watershed.

The most appropriate method to determine the distribution is from data collected in the watershed. In the literature, many different distributions have been used, including normal, log-normal, uniform linear (linear on both sides of the median value), Pareto, Poisson, Gaussian, Weibull, and many more. However, in the absence of data, a theoretical distribution may be used. For example, hydraulic conductivity in aquifers has been demonstrated to exhibit a log-normal distribution. For a Monte Carlo analysis, choosing the type and shape of a distribution is one of the most challenging aspects. Please refer to the paper by Zhang (2002).

10.4.3.5 Monte Carlo Analysis in Spreadsheet Models

Monte Carlo simulations can be useful in simple spreadsheet models, particularly when used for screening and preliminary decision making. A similar process must be followed as described above for obtaining input-parameter distributions for lumped-parameter models. Then, the analysis of model output with respect to decision making is similar to that described above in Section 10.C.

Two excellent resources for using Monte Carlo simulations in Excel™ spreadsheets are provided by the websites below. These websites also include some basic theory on Monte Carlo simulations.

www.Solver.com

www.riskamp.com

www.treeplan.com

CHAPTER 11.0

CASE STUDIES

11.1 Example for Following User's Guide for Watersheds

11.1.1 Evaluating Perceived Watershed/OWS Problem

If onsite systems are not properly sited or installed, or if placed in soils that do not have adequate capacity for wastewater renovation, they can be a potential source of pollution. A comprehensive watershed based approach should be implemented if there are perceived problems associated with onsite systems as a result of new development or population growth. These problems could be increased concentration levels of OWS pollutants such as nutrients, emerging organic chemicals (EOCs) including pharmaceuticals and personal care products (PPCPs) in stream or groundwater.

Information on the trends of new developments involving OWS can indirectly imply that there could be potential water quality problem but this should also be supported by water quality monitoring. In the development of the user's guide the researchers selected two watersheds; the Turkey Creek watershed (TCW) in Colorado and the upper Deschutes watershed (UDW) in Oregon, with OWS pollution concerns primarily due to increasing population levels. These case studies demonstrate how OWS can be modeled at a watershed scale.

11.1.2 Applying Initial Screening Model

Various modeling tools are available for evaluating pollutant movement from land sources to waterbodies. Models range from simple regression models that relate nutrient loads to basin characteristics, to the complex physically based models that account for all of the known processes for simulating the water cycle and nutrient fate based on climate, topography, soil properties, land use, and management practices.

The screening models describe the system without including the detailed processes involved or data density that would be necessary for site-specific delineations. They can be mass balance models or GIS-based models. A review of screening models is included in the Appendix A. Application of simple screening models can save time and money required for setting up and running more complex models. They provide cost-effective evaluations on strategies before more expensive physically based models are employed for a specific site. A screening model can be used to assess the problem before a complex model is used.

If a simple mass-balance modeling shows that pollutant concentrations in an aquifer or stream are below the desired limit using conservative assumptions, then additional complex modeling may not be required. If the screening model indicates a problem, one can use a more complex numerical model. A GIS based screening model is applied to one of our watersheds (UDW). A physically based watershed model is applied to both watersheds, UDW and TCW. A groundwater modeling was applied to the UDW.

11.1.3 Selecting Type of Model

Once a problem is identified through monitoring or a screening model and a decision is made to implement a more complex model, the next step is to choose the right type of complex models. Model selection for OWS requires evaluation of key features of the models and ability to handle non-point source pollution from OWS.

Emphasis has to be given to models that can simulate the most common wastewater pollutants, especially nitrogen and phosphates both at a field and watershed scale. Sediment transport should also be considered since transport and fate of sediments and nutrients, particularly phosphorus, are intimately related.

The model type should be chosen depending on the type of problem, whether it is a ground or surface water pollution problem. The models included in the initial review include AGNPS, ANNAGNPS, ANSWERS-2000, CREAMS-WT, GLEAMS, HSPF, MIKE-SHE, SWAT, MODFLOW, SWMM, WARMF, and WMS.

In general, models such AGNPS, ANSWERS-2000, ANNAGNPS, CREAMS-Wt, GLEAMS, SWAT and HSPF have similar routines for nutrient transformation. Nitrogen, phosphorus, and pesticides in these models are based on routines developed for the CREAMS/GLEAMS models including biochemical processes and groundwater loading. However none of these models explicitly account for OWS.

AGNPS and ANSWERS-2000 are primarily surface runoff models and do not handle subsurface flow well, therefore are not suited to simulate OWS pollutants. Models like ANNAGNPS, CREAMS-WT and GLEAMS have routines to simulate subsurface flow and leaching but subsurface flow and leaching do not contribute to stream flows. Although these models can be good to simulate the effect of OWS pollutants on groundwater, they are not suited to simulate the impact of OWS pollutants on stream flow. On the other hand, models like SWAT and HSPF have routines to simulate lateral and groundwater contributions to stream flow. Although these models don't explicitly account for OWS pollutants, they can still be used using the non-point and point source routines features of the models.

Compared to WARMF, HSPF and SWAT model do not explicitly account for OWS. Detailed case studies using WARMF are therefore included in this guide.

The three dimensional MODFLOW groundwater flow model from the US Geological Survey has been used along with GIS to quantify septic system nitrogen loadings to receiving waters (Sham et al., 1995). Morgan and Everett (2005) used MODFLOW-MT3D in conjunction with optimization model to estimate the optimal loading of nitrate from decentralized wastewater treatment systems to an aquifer. This study is also included in the guide.

11.1.4 Obtaining Data

Obtaining and organizing data is a major part of developing a successful watershed model. It is often the most time consuming task in watershed modeling. Most models may require information on soils, topography, and land use. Again after sub watersheds and land uses are defined, input default parameter values need to be modified to reflect spatial information within each subwatershed. For most models, modifying these parameters depends on the soils, topography, and land use. For example, for phosphorus transport the adsorption isotherm should be defined based on soil type. If such information is not available, the parameter can be obtained via calibration.

Although the data-gathering and analysis phases of the watershed planning process are very important in estimating source loads, they can also be very challenging. The types and amount of data available vary by watershed, and there is often a variety of data, making it difficult to decide which data are useful. These can also affect the decision on the type of model to use because models also vary with respect to input data requirement.

The methods of obtaining data and the data sources are presented in Chapters 8.0 and 9.0. This section presents the methods and the data sources for the two watersheds included in the case study.

Turkey Creek Watershed (TCW): The 126 square-kilometer TCW is located in Jefferson County, approximately 30 km southwest of Denver, Colorado.

A Digital Elevation Model (DEM) from the USGS database defined the topography. Land-use data were derived from the 1992 National Land Cover Dataset (NLCD). Soil type was determined using STATSGO data (USDA, 2007). Meteorological data were obtained from National Climatic Data Center (NCDC) (NCDC, 2007). (The digital elevation model (DEM), the land use map and the soils map were presented in Chapter 7.0.)

The topography is mostly steep, with elevations ranging from about 1800 m to 3200 m. Agricultural land accounts for only 0.35% of TCW, thus crop rotation from year to year is not significant in TCW. Bossong et al. (2003) estimate that forest canopy covers 60-70% of TCW, while the 1992 NLCD data shows that forest covers about 67% of TCW. Thus, there has been no considerable land use change in the watershed in recent years.

WARMF model accepts the number of people per catchment to calculate the septic loading. The data can be obtained from census data. However the most recent census data on OWS distribution is from 1990. The recently released 2000 data does not include a geographic distribution of septic systems at the watershed scale, so we determined geographic distribution of OWS using county tax records, which indicate septic systems installed at a residence. These data were obtained from the Environmental Health Services Division, Jefferson County Department of Health and Environment. The septic data has a GIS shape file with locations of septic systems that existed in 2000 with points located at the centroid of each tax lot. The septic data was then assigned to each catchment in the watershed using arcview's geoprocesing spatial join tool. The septic system distribution over the TCW is shown in Figure 11-1.

The total number of residential lots in TCW is 4363 and assuming 2.5 persons per lot, the total population in the watershed is about 11,000. There are five wastewater treatment plants (WWTP) in the TCW. These include Conifer Metro District, Aspen Park Metro District, Conifer Sanitation District, Conifer High School and WWTP at Tiny Town. Some of them (Conifer and Aspen Metro districts) use subsurface infiltration gallery discharge while others (Conifer sanitation district, Conifer high school and Tiny Town WWTP discharge into surface water. The WWTPs were added to WARMF as point sources. Thus, almost all the residential dwellings in the watershed use septic systems. The septic systems are treated as non-point sources in WARMF while the WWTPs are treated as point sources.

Stream flow data and water quality data are required for hydrologic and water quality calibration. Stream discharge data from two USGS stream gaging stations were used to calibrate the WARMF model. The two stations have USGS gage numbers 06710995 (old station) and 06710992 (new station), on TCW at the mouth of the canyon near Morrison and TCW near Indian Hills, respectively. Station 06710995 operated from April 1998 to April 2001 and was discontinued in lieu of the new, up-graded station, which operated from April 2001 to 2007.

To calibrate water quality predictions, water quality data are needed. This data can be obtained from government sources or other local sources. Water quality data collected within flow calibration period 1998 through 2003 were used to guide the water quality calibration. The data was obtained from United States Geological Survey site (USGS, 2001) and from a report on investigation of the fate of individual sewage disposal system effluent in TCW, Colorado (Dano et al., 2004).

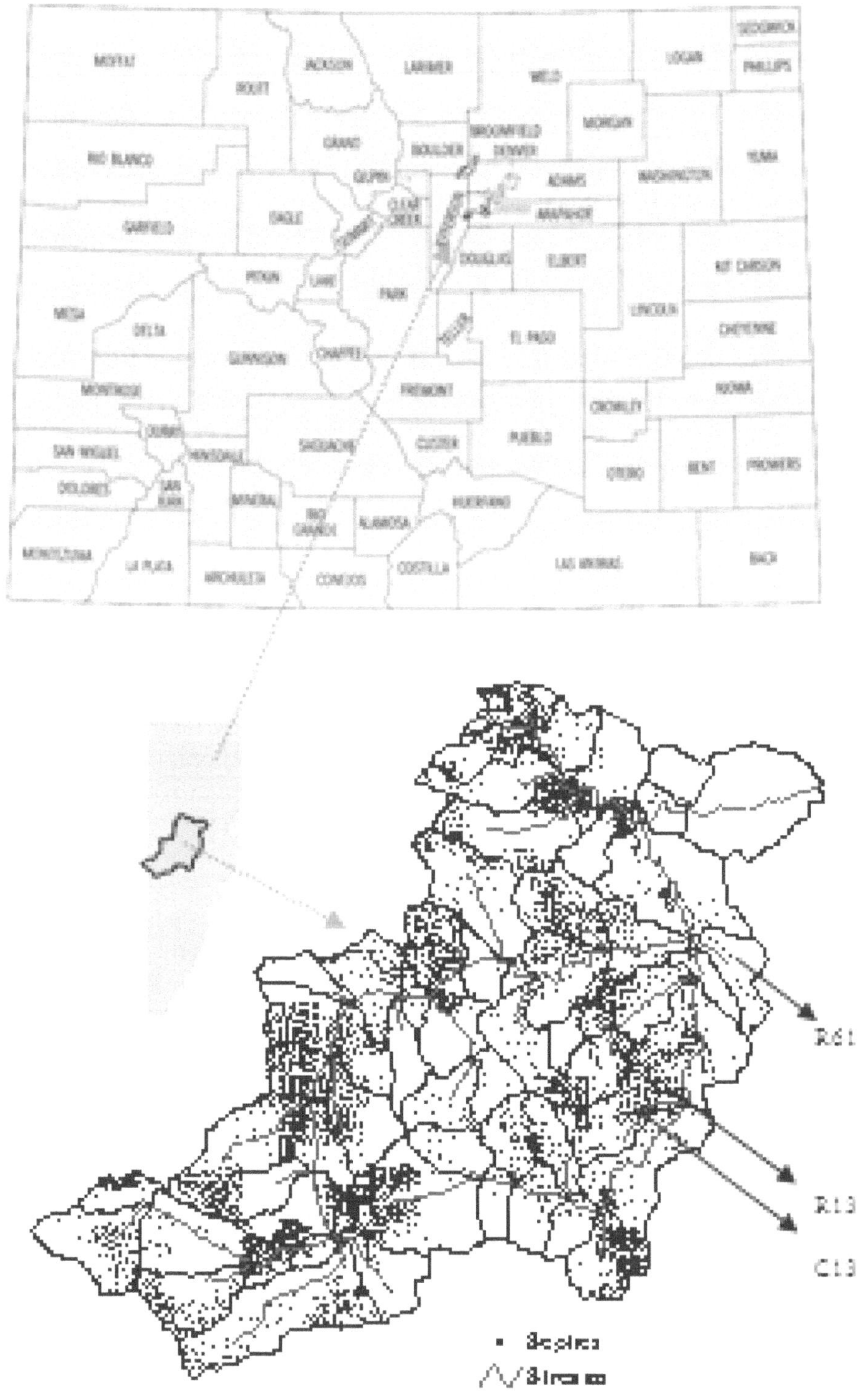

Figure 11-1. OWS Distribution in Turkey Creek Watershed, Colorado.

Upper Deschutes Watershed (UDW): The 4525 square-kilometer UDW is situated in Central Oregon. The majority of the watershed is in Deschutes County with smaller portions in Klamath County. The topography is mostly steep with elevations ranging from 1200 m to 3200 m.

A Digital Elevation Model (DEM) from the USGS database was used to define the topography (Figure 11-2). Land-use data were derived from the 1992 National Land Cover Dataset (NLCD). Based on this land use data set, forest covers about 80% of watershed. The remaining 20% accounts for grass, shrubs, lakes and few agricultural sites. Agricultural land accounts for less than 2% of watershed.

The climatic data was obtained from Oregon Climate Service (OCS). The OCS has precipitation and temperature data for several stations from mid 1950s to present. The metrological stations are Wickiup Dam station (359316), Bend (350694), Brothers (351067), Madras (355139), Redmond FAA AP (357062), Chemult (351546) and Odell Lake east (356252), with elevations ranging from 680 m to 1500 m.

Additional data required to construct WARMF model such as wind speed, air pressure, and cloud cover were also collected from National Climatic Data Center (NCDC) (NCDC, 2007). The watershed has several lakes. The GIS shape files for the lakes were obtained through the National Hydrographic Database (NHD). The NHD is a comprehensive set of digital spatial data that contains information about surface water features such as lakes, ponds, streams, rivers, springs and wells. Data on lake geometry such as stage-area and stage discharge (rating curve) for reservoirs has also obtained from the OCS.

Soils in the upper Deschutes river subbasin are formed partially to entirely from materials deposited by volcanic eruptions. The STASTGO soil information was used in this study. Based on STATSGO soil data, there are nine soil categories in the watershed.

The geographic distribution of OWS/septic data was based on GIS data on Deschutes county tax lot records. The GIS shape file with locations of septic systems located at the centroid of each tax lot was used. This information was used to assign the number of people using septic systems to each catchment in Deschutes watershed, as shown in Figure 11-3. Based on the GIS data, the total number of residential lots in Deschutes watershed is 5185 and assuming 2.5 persons per lot, the total population is about 12,963.

The watershed was delineated with an outlet at the USGS gage 14064500 on the Deschutes River at Benham Falls, near Bend. Two stream gages were used to calibrate the watershed model: the gage station 14064600 located at Deschutes River below Benham Falls near Bend and another gage station upstream (14056500) located at Deschutes River below Wickiup reservoir near Lapine. Historical data for these gage stations were obtained from Oregon water resources department, a public source. Water quality data collected within flow calibration period 1996 through 2005 were used to guide the water quality calibration. Total phosphorus and nitrate concentration data was also obtained from the same location: Harper's Bridge location shown in Figure 11-3, from Oregon Department of Water Quality (DEQ) website.

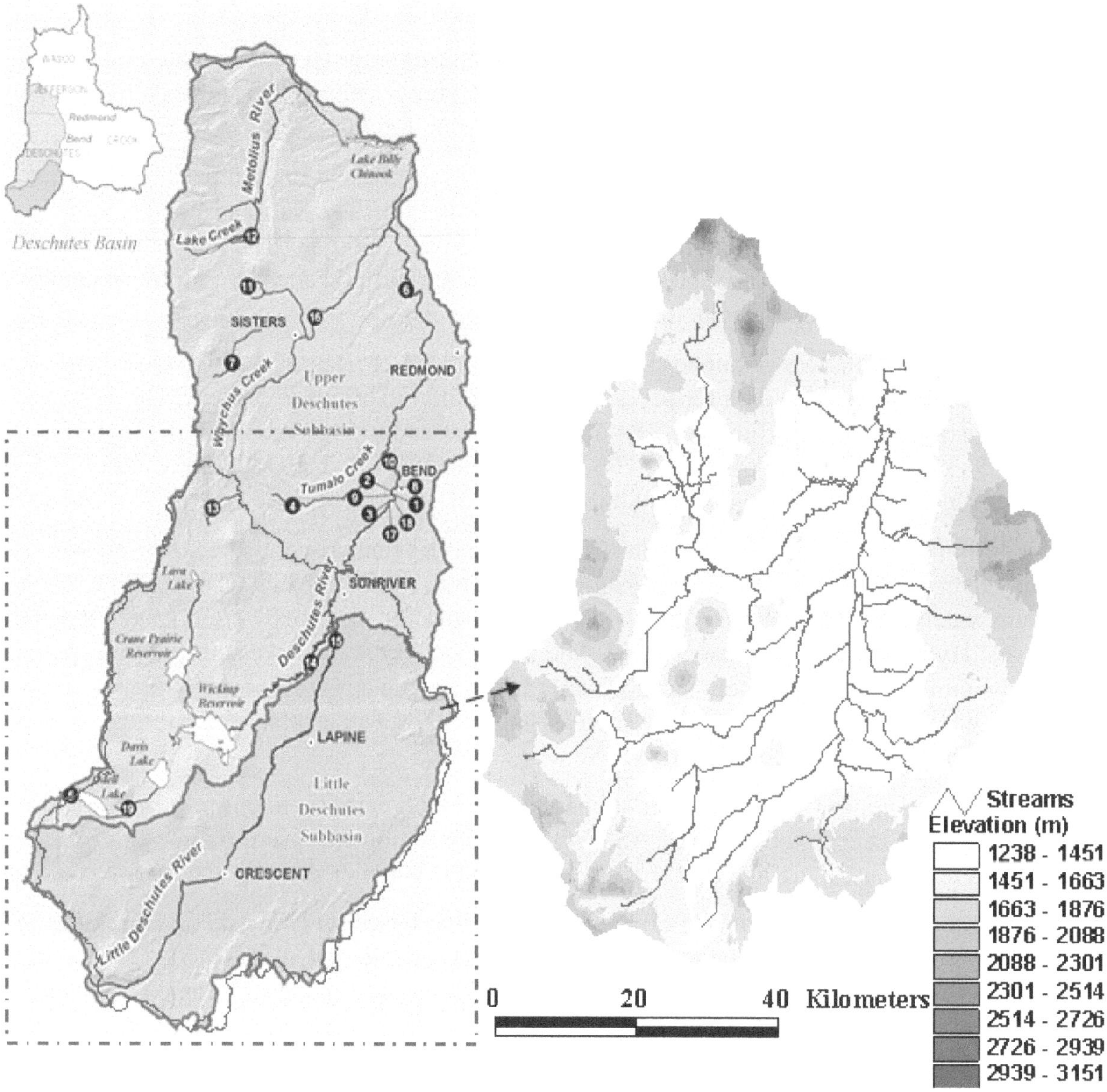

Figure 11-2. Location of Upper and Middle Deschutes River Watershed, Oregon.

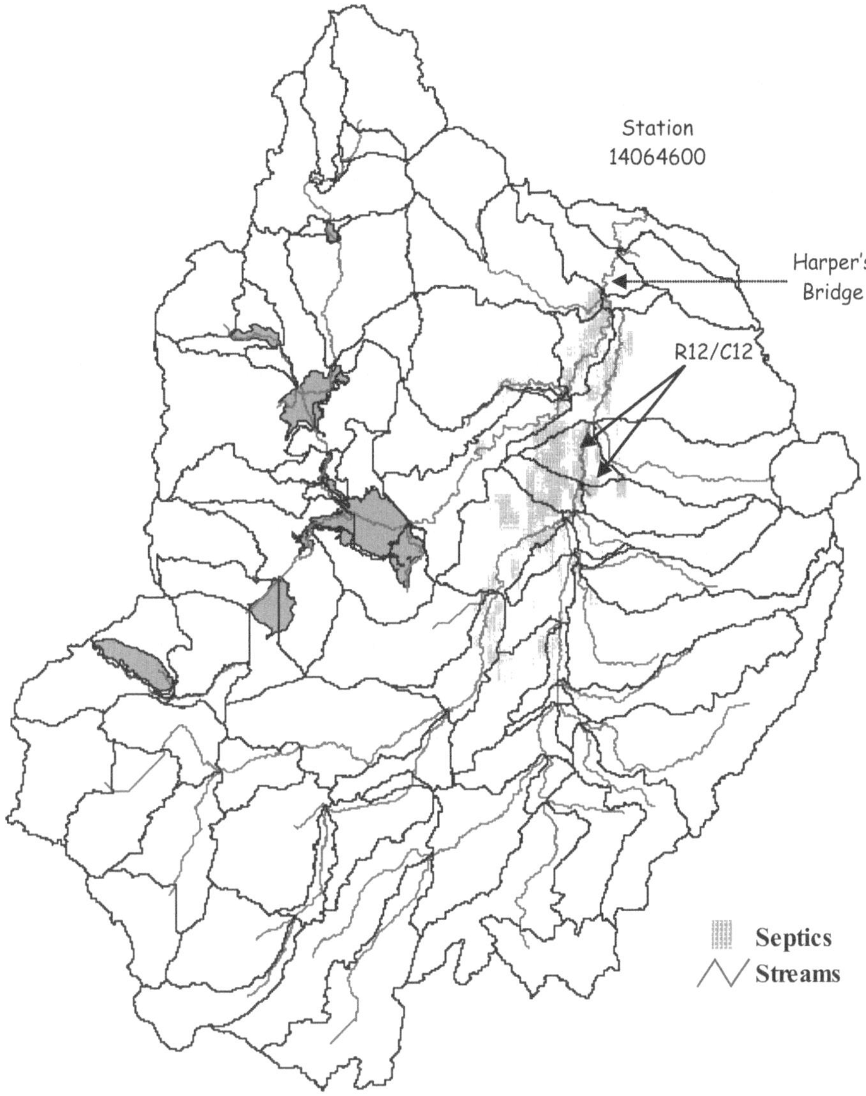

Figure 11-3. Locations of OWS, Stream Gages, Sediment, and Nutrient Observation Points.

11.1.5 Assessing Data Needs

Even after setting up the watershed model, the data at hand may not be sufficient. Additional data may be needed to completely characterize the watershed, especially during the calibration stage. If data already available is not sufficient, it is necessary to collect additional data during the planning phase for complete characterization.

Review your data to deterimine its adequacy and quality. This will help you decide whether to collect additional data before proceeding with data analysis. Data review will help identify major gaps and then determine the quality of the data. Obvious data gaps can be identified during the data inventory process, but more specific data needs are often discovered only during data analysis and subsequent activities. The data gap could be spatial or temporal. Sometimes it may happen that there is a spatial data gap for the data that were collected within the time frames of interest or the data may have a temporal gap. Available data may be too old collected when watershed conditions were very different, reducing the importance of the data to the current situation.

11.1.6 Calibrating a Model

Models are applied to an existing condition to set a baseline for comparison. The ultimate goal of modeling is to make predictions for scenarios that are different from the existing conditions. A prediction may involve some future land use, management practice or point source implementation alternatives, or the effect of new development or population growth on stream and groundwater quality, etc.

Model calibration should be based on existing conditions before being used for predictions. Existing conditions reflect the data used to build the model. Modeling results need a reality check before they are used to support a loading analysis or evaluation of management scenarios. The model should be calibrated to ensure accurate representation of the watershed processes. If a model is designed to evaluate annual nutrient loads, comparisons are made with flow and nutrient monitoring information. The calibration process should focus on the questions the model is designed to answer.

When data are limited, the model user may compare model results to literature values or data from surrounding watersheds. Sometimes, when data are highly limited, model testing is based primarily on comparison with literature values, similar studies in nearby regions, and evaluation using alternative calculation techniques.

In cases where additional data gathering is not possible and historical records are limited, calibration might be based on a single downstream location. Calibration is best performed at locations where flow gauging and water quality sampling are available, typically at USGS gauging stations.

When selecting the subwatershed delineation in the initial model setup, it is important to consider the locations of available monitoring and testing points. Then the model output can be compared at the locations where flow and water quality measurements are available.

The calibration process usually flows a hierarchy with flow checked first, followed by sediment, and then the various pollutants (e.g., nutrients). Generally, calibration may involve checking water balance, observed versus measured flow (daily average, monthly, annual, and flow duration curves), checking observed versus measured load (annual loads, seasonal variation, source loads) or observed versus modeled pollutant concentrations.

The researchers calibrated their watershed for hydrology, sediment and nutrients before using them for scenario evaluation. For our watersheds, hydrology and water quality calibration were done using automated calibration software called UCODE. Details about autocaibration using UCODE are presented in Poeter et al. (2005).

11.1.7 Model Evaluation

It is important to know how good the calibration results are and how to measure the goodness of fit between measured and simulated values. Both statistical and graphical model evaluation techniques are applied in calibration and validation of models. The statistical methods include standard regression, dimensionless and error index. Graphical methods can also be used to visually compare measured and observed values to get an overview of model performance.

Details about model evaluation are presented in Chapter 9.0.

Water quality monitoring data can be used to verify a calibrated model. Watershed modeling and monitoring activities should be better integrated to reveal key uncertainties, provide focus for strategic research, and improve model accuracy and interpretation of the observations. This integration involves both observations in the environment (monitoring). Simpler models may require fewer monitoring data relative to more complex models that have great data needs.

11.1.8 Model Implementation

A well-calibrated model can be used for prediction. Scenarios can be evaluated using the calibrated model. The researchers used the calibrated model to evaluate OWS related scenarios such as population growth, conversion of onsite septic systems to public sewers and effect of septic system discharge concentrations. The results are summarized in the following section.

11.2 Scenario Simulations for Potential OWS-Related Issues in TCW and UDW

Details of each case study in this section can be accessed via the information provided in Appendix C. We present below a summary of case studies on nitrate loading to groundwater from OWS near La Pine, Oregon, and an evaluation of scenarios related to OWS in TCW, Colorado and UDW, Oregon. Another case study on OWS (Heatwole and McCray, 2007) for a different location is referenced in Appendix C.

11.2.1 Nitrate Loading to Groundwater from OWS Near La Pine, Oregon

The population of rural residential areas near La Pine in southern Deschutes County and northern Klamath County, Oregon, has grown rapidly since the 1960s. Most of these areas lie within a tract adjacent to the Deschutes and Little Deschutes Rivers that extends roughly 25 miles south of Sun River (see Figure 11-4). Existing and future homes on more than 9,300 residential lots in the area currently use, or will use, individual, onsite septic systems for wastewater disposal and shallow wells for water supply. At least 50% of these wells draw groundwater from the upper 50 feet of the shallow aquifer that underlies the area (Morgan et al., 2007).

Vulnerability of the shallow aquifer to contamination has led to concern by residents, county planners, and state regulators. They are concerned that wastewater from septic systems may pose a threat to the primary drinking water supply if residential development continues at planned densities using conventional septic systems. Another concern is the quality of local streams (Hinkle et al., 2007). The Deschutes and Little Deschutes Rivers, which flow through the developed areas near La Pine, already have excessive algae in some reaches, possibly due to nutrient (nitrogen and phosphorus) contributions from groundwater.

Conventional residential septic systems are the principal source of nitrogen to the shallow aquifer in the La Pine area (Hinkle, et al., 2007), and the nitrate contribution (loading) to the aquifer from these septic systems has increased rapidly as a result of ongoing residential development. The U.S. Environmental Protection Agency has established 10 parts per million (ppm) of nitrogen as the maximum allowable nitrate concentration in drinking water for public water supply systems. Oregon law sets a nitrate concentration of 7 ppm as the level at which regulatory action must be taken to control water quality degradation.

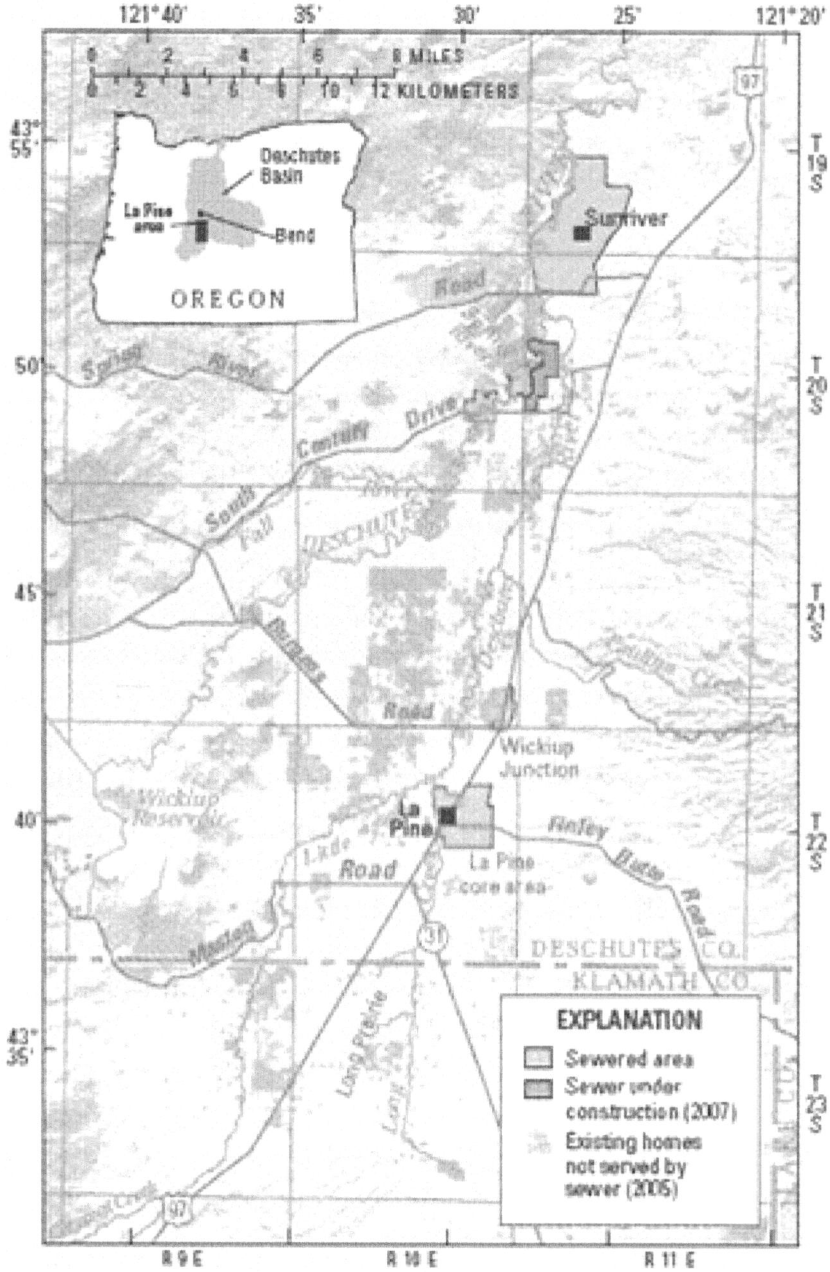

Figure 11-4. Map of La Pine Study Area Showing Sewered Areas and Areas Served by OWS (from USGS Fact Sheet 2007-3103).

Computer models simulate the physical and chemical processes affecting the fate of nitrate in the shallow aquifer system near La Pine. The purpose of the models was to: 1) test concepts of how hydrogeologic and geochemical processes affect the movement of nitrate through the groundwater system; and 2) provide tools that could evaluate future water-quality conditions and alternative management options.

The U.S. EPA has awarded grant funding to Deschutes County for a project to protect the drinking water resource in south Deschutes County by using advanced wastewater treatment systems. Using extensive data and state-of-the-art models developed during the La Pine National

Demonstration Project, Deschutes County will develop treatment standards for wastewater treatment systems. It will also establish financial incentive programs to help residents offset the cost of protecting and improving the region's water quality. Financial incentives may be in the form of rebates or low interest loans depending on the source and quantity of funds.

Owing to low recharge, groundwater moves slowly through the shallow aquifer. Because groundwater moves slowly, it takes a long time for nitrate to appear in well water. Most wells currently provide drinking water that percolated to the water table decades ago, when there were very few homes and septic systems. Nitrate plumes, however, are beginning to affect a significant number of drinking water wells. Much of the nitrate in the aquifer currently is confined to plumes less than about 30 feet below the water table, so not all supply wells are drawing water from affected areas of the aquifer. As development proceeds and the nitrate plumes expand and move deeper into the aquifer, more wells will be affected.

The hydrogeologic framework for the aquifer system was modeled using available well data with geostatistical methods. The framework consisted of a three dimensional representation of the distribution of major sediment types (clay, silt, sand, gravel).

Hydraulic conductivity was estimated using information available on reports filed by drillers. Initial values of porosity and dispersion coefficients were estimated using values reported in the literature for similar materials. All three parameters were adjusted to some extent during model calibration until simulated hydrologic and water quality conditions matched measured conditions. Some parameter values, such as recharge, were available from previous studies in the area.

The capacity of the aquifer to receive nitrate varies throughout the area and depends on factors related to geology, climate, chemistry, and nearby development. These factors are accounted for by the model, allowing it to compute the maximum sustainable nitrate loading capacity in subareas. The maximum sustainable loading capacity also depends on the water quality protection goals for the aquifer. Model users set the values of water quality goals, which can be the maximum acceptable nitrate concentration in groundwater, the maximum acceptable discharge of nitrate to streams, or both. The model can be used to examine the trade-offs between more stringent water quality goals and the costs of limiting nitrate loading. Planners and resource managers also can use the model to identify areas where loading from planned or existing development exceeds the sustainable nitrate loading capacity of the aquifer and devise appropriate strategies for reducing loading.

11.2.2 Scenarios Related to OWS

11.2.2.1 General Watershed Descriptions

OWS is becoming a concern in areas with rapid development and population growth. Figures 11-1 and 11-3 show the study areas and existing distribution of OWS in TCW and UDW. The TCW is representative of mountain watersheds along Colorado's Front Range. The basin is a topographically defined watershed in the foothills of the Front Range, approximately 35 km (22 miles) southwest of Denver. The basin lies completely in Jefferson County and covers 122 km^2 (47 mi^2). It includes the mountain towns of Aspen Park, Conifer and Indian Hills.

Recent development in the watershed has produced noticeable effects on the quality of the surface and groundwater. Bossong et al. (2003), Morgan (2000), Hofstra and Hall (1975), and Yacob (2004) report that most chemical constituents in both the ground and surface water increased from 1975 to 1999. With the increase of the Denver metro area's population, the number of people living in the foothills west of Denver also increased. The majority of people living in the Jefferson County foothills rely on individual domestic wells for drinking water and septic systems for the disposal of wastewater.

The central Oregon community of La Pine is a rapidly growing rural residential area without centralized wastewater treatment or drinking water systems (Figure 11-4). Most homes rely on individual decentralized wastewater treatment systems for wastewater disposal and wells for water supply. Wells are typically shallow (less than 50 feet below land surface) to tap permeable sands and gravels and to avoid more mineralized groundwater found in deeper aquifers. The water table is also shallow (less than 10 feet below land surface), and thin volcanic soils provide little opportunity for removal of nitrogen before septic effluent recharges groundwater. Centralized sewer or water systems have been determined to be economically infeasible in the area (KCM Inc., 1997) and, with a large number of lots still available for development, planners and regulators are concerned that future growth will render the groundwater resource unusable. The Deschutes and Little Deschutes Rivers have been listed as water quality impaired under Section 303(d) of the Clean Water Act. High nitrate concentrations in shallow groundwater have been identified as potential contributors to eutrophic conditions found in some reaches (Anderson, 2000 and Jones, 2003).

Selection of models that can be applied to OWS requires evaluation of key features of the models and an ability to handle non-point source pollution from OWS. Models like SWAT and HSPF have routines to simulate lateral and groundwater contributions to stream flow. Although these models don't explicitly account for OWS pollutants, they can still be used with the non-point and point source routines features of the models. Compared to WARMF, HSPF and SWAT model do not explicitly account for OWS. Thus, WARMF was used for the case studies that follow because WARMF has a routine that accounts for OWS pollutants. WARMF has limited groundwater capabilities.

11.2.2.2 Phosphorus Related Scenarios

Phosphorus is considered to be the nutrient most responsible for eutrophication. Nutrients are usually associated with agricultural sources; however, other sources such as atmospheric deposition and subsequent wash off from impervious surfaces, and septic system effluent, are also becoming a concern.

Between 20 and 30% of total phosphorus in raw wastewater is separated out in the form of sludge in a septic tank (Wood, 1993). Orthophosphate may also be removed in septic tanks through mineral precipitation reactions (Zanini et al., 1998). In a soil absorption system, septic tank effluent is discharged via a dispersal system to the soil infiltration zone, the vadose zone, and, ultimately, to groundwater. Most of the phosphorus and pathogen removal occurs in the vadose zone (U.S. EPA, 2002).

Most research has shown that phosphorus plumes are unusual in unsaturated soils with finer textures because most phosphorus is absorbed by the soil (Stolt and Reneau, 1991), and numerous studies have documented a high degree of phosphorus removal within the first few meters down gradient from the drain field (Weiskel and Howes, 1992; Wilhelm et al., 1994). In properly functioning systems not located in soils conducive to plume formation, as much as 95% of the phosphorus may be retained in the soil (Mandel and Haith, 1992). A modeling study on Phosphorus Transport in the Blue River Watershed, Summit County, Colorado, using the SWAT model, indicated that OWS are not the primary source of P in the lake. Instead, P in runoff sediments is the most likely contributor to surface water (Lemonds and McCray, 2003).

Different assumptions are used in the design of onsite systems regarding the movement of phosphorus from septic systems to waterbodies. Some methods assume all available onsite phosphorus from the watershed reaches waterbodies, others assume a fraction of the phosphorus from septic effluent reaches waterbodies without accounting for detailed processes. Yet others assume that only phosphorus from within an assumed distance from waterbodies reaches the waterbodies. In order to understand phosphorus removal in soil absorption systems, basic phosphorus geochemistry and the physical processes should be known. Also, for P transport in surface runoff, controlling sediment loading reduces P loading into the stream. Thus, assumptions that all or part of P applied reaches surface or groundwater without considering site conditions may lead to invalid results.

This study evaluated the impact of different scenarios related to OWS on phosphorus concentration both in stream and in soil water. These scenarios are discussed below.

Impact of Population Growth - Turkey Creek Watershed One common source of concern is an increase in the number of people using septic systems and the resulting change in water quality. A calibrated model was used to evaluate the effect of population growth. This scenario evaluated the impact of residential development and an increase in the number of people using septic tanks on stream and soil water P concentration. The total number of residential lots in TCW is 4,363. Assuming 2.5 persons per lot, the total population in the watershed is about 11,000. The population of Turkey Creek and the area under residential houses was doubled.

The increase in average stream P concentration of 4.3 and 5.1% respectively for the two stream segments, R61 and R13 (Figure 11-1) are shown in Table 11-1 below. The soil water total P concentration was also evaluated at catchment C13. The total number of houses in the catchment was 200. The increase in soil water P concentration for the bottom soil layer for this catchment was 0.95%.

Impact of Population Growth - Upper Deschutes Watershed An identical population growth scenario, as in TCW, was evaluated in the UDW using GIS data and tax lot records for Deschutes County. Based on the GIS data, the total number of residential lots is 5,185 and, assuming 2.5 persons per lot, the total population is about 12,963. This population growth scenario involved evaluation of the impact of residential development and increase in the number of people using septic tanks on stream and soil water P concentration. The population using septic systems was increased by 100% from the base population. The residential area was also increased in the same proportion.

Total P concentration was evaluated at two stream segments shown in Figure 11-3, namely, at the stream segment near Harper's bridge where phosphorus was calibrated and at the river mile 13.7 upstream from the intersection of Paulina Creek and the Little Deschutes River (R12). Total P concentration was also evaluated for soil water for catchment 12 (C12) shown in Figure 11-3 upstream of river mile 13.7. The catchment has about 800 houses. The results show that there is no increase in average total P concentration for the two stream segments. The soil water total P concentration for the bottom soil layer increased 0.09% for an increase of 100%. The results are summarized in Table 11-1.

The increase in average stream P concentration for the two river segments in TCW when the population was increased by 100% was 4-5%. There was no change in average stream P concentration for stream segments in UDW. The discrepancies are attributed to the differences in flow rates. The average simulated flow rate was 34 m^3/s for the Deschutes River at Harper's bridge, thus there was no impact from septic systems due to the effect of dilution compared to an average stream flow of 0.42 m^3/s for TCW. The impacts of septic systems on total P concentration in soil water are generally low.

Table 11-1. Effect of Population Growth on Total P Concentration.

Watershed	Location	Scenario	Total P concentration (mg/l)	% change
TCW	Stream Segment-R61	Base Scenario	0.02103	--
		Scenario 1	0.02194	4.31
	Stream Segment R13	Base Scenario	0.02880	--
		Scenario 1	0.03029	5.15
	Bottom soil layer (C13)	Base Scenario	0.05240	--
		Scenario 1	0.05290	0.95
UDW	Stream Segment-Harper's Bridge	Base Scenario	0.06860	--
		Scenario 1	0.06860	0.00
	Stream Segment R12	Base Scenario	0.08390	--
		Scenario 1	0.08390	0.00
	Bottom soil layer (C12)	Base Scenario	0.08513	--
		Scenario 1	0.08520	0.09

Impact of Effluent Concentration - Turkey Creek Watershed This scenario involved increased loading as a result high effluent concentration from septic tank. Total P concentrations were compared for the base scenario and a new scenario with a higher effluent concentration. McCray et al. (2005) developed cumulative frequency distributions (CFDs) for OWS effluent concentrations of N and P based on data gathered from existing studies reported in the literature. For the base scenario, a median (50%) value (10 mg/l) from the CFDs was used. For the prediction scenario, concentration of 22 mg/l representing 100% cumulative frequency values (maximum values reported) was used. The results are summarized in Table 11-2. The increase in average stream P concentration for the two stream segments, R61 and R13, for increased effluent concentration level was 4.41 and 5.81 % respectively. The soil water total P concentration evaluated at catchment C13 water P concentration for the bottom soil layer increased by 0.95%.

Impact of Effluent Concentration - Upper Deschutes Watershed The same scenario was evaluated for UDW. Total P concentration was evaluated at two stream segments (stream segment at Harper's bridge and stream segment R12) and soil water for catchment C12. The results, summarized in Table 11-2, show no significant change in average total P concentration for the two stream segments. The increase in soil water total P concentration for the bottom soil layer was also low but relatively higher than the changes in the stream concentration values. The relatively lower impact on stream concentration compared to TCW is attributed to the dilution in UDW due to the relatively higher flow rate.

Table 11-2. Effect of Septic Tank Ffluent Concentration.

Watershed	Location	Scenario	Percentile	Effluent conc. (mg/l)	Total P conc. (mg/l)	% change
TCW	Stream Segment-R61	Base scenario	50	10.0	0.02103	--
		Scenario 2	100	22.0	0.021958	4.41
	Stream Segment R13	Base scenario	50	10.0	0.02103	--
		Scenario 2	100	22.0	0.03047	5.81
	Bottom soil layer (C13)	Base scenario	50	10.0	0.0524	--
		Scenario 2	100	22.0	0.0529	0.95
UDW	Stream Segment-Harper's Bridge	Base scenario	50	10.0	0.06861	--
		Scenario 2	100	22.0	0.06863	0.023
	Stream Segment R12	Base scenario	50	10.0	0.08398	--
		Scenario 2	100	22.0	0.08401	0.032
	Bottom soil layer (C12)	Base scenario	50	10.0	0.08513	--
		Scenario 2	100	22.0	0.08521	0.097

Impact of Conversion of OWS to Conventional Sewers -Turkey Creek Watershed The model was used to evaluate the effect of conversion of onsite systems to public sewer systems. Two scenarios were evaluated. The first scenario (a 50% conversion) involved conversion of 50% of the population using septic systems to sewers, and the second scenario (a 100% conversion) involved a 100% conversion of existing OWS to sewers. The increase in average stream P concentration as a result of conversion was 13.1% and 5.6% for stream segment R61 and R13 respectively when 50% of the OWS are converted to sewer. The corresponding percentage increase in the average stream P concentration when all OWS are converted to sewer use was 26% and 11.4% for stream segments R61 and R13, respectively. There was relatively small reduction in the soil water concentration for bottom soil layer as a result of conversion to sewers. The results are shown in Table 11-3. This indicates that, depending on the treatment efficiency, WWTPs may increase stream P concentration and lower soil water concentrations.

Impact of Conversion of OWS to Conventional Sewers - Upper Deschutes Watershed There was also an increase in stream P concentration when septic systems were converted to sewers in the UDW, but the increase was small relative to TCW (shown in Table 11-3). The increase in average stream P concentration as a result of conversion was 0.2% and 0.41% for stream segment at Harper's bridge and stream segment R12 respectively when 50% of the population was converted to sewer use. The corresponding percentage increase in the average stream P concentration when all OWS are converted to sewer was 0.21 and 0.43%, for stream segment at

Harper's bridge and stream segment R12, respectively. There was a small reduction in the soil water concentration for bottom soil layer as a result of conversion to sewers. This indicates that, depending on the treatment efficiency, WWTPs may increase stream P concentration and decrease soil water concentrations.

Table 11-3. Impact of Conversion of OWS to Conventional Sewers on Total P Concentration.

Watershed	Location	Scenario	Total P concentration (mg/l)	% change
TCW	Stream Segment-R61	Base Scenario	0.02103	—
		Scenario A	0.02379	13.15
		Scenario B	0.02662	26.56
	Stream Segment-R13	Base Scenario	0.02880	—
		Scenario A	0.03046	5.76
		Scenario B	0.03209	11.42
	Bottom soil layer (C13)	Base Scenario	0.05240	—
		Scenario A	0.05210	-0.57
		Scenario B	0.05170	-1.33
UDW	Stream Segment-Harper's Bridge	Base Scenario	0.06860	—
		Scenario A	0.06880	0.20
		Scenario B	0.06890	0.41
	Stream Segment R12	Base Scenario	0.08390	—
		Scenario A	0.08420	0.21
		Scenario B	0.08430	0.43
	Bottom soil layer (C12)	Base Scenario	0.08513	—
		Scenario A	0.08517	-0.015
		Scenario B	0.08520	-0.047

Impact of Soil Adsorption Capacity - Turkey Creek Watershed Parameter sensitivity analysis on the watershed showed that adsorption is an important parameter for estimating stream and soil water P concentration. This scenario was implemented in TCW to evaluate the effect of reduced adsorption. This can be related to the cases where phosphate-adsorption capacity becomes depleted leading to more phosphorus being carried into nearby streams. If the adsorption capacity of soils to retain phosphorus is reached (as predicted by isotherms), soils may reach saturation capacity and phosphorus may be exported off-site into surface and groundwater. Phosphate adsorption capacity may be significantly lower in coarser grained soils as in TCW.

The modeling study showed that for a conventional septic tank with effluent P concentration of 10mg/l, if adsorption capacity is reduced from a calibrated value of 14.37 L/kg, the stream P concentration increases considerably and would increase by 400% if the adsorption capacity is totally depleted (0 L/kg). Thus, adsorption capacity of the soil is even more important than the scenarios presented earlier, population growth or increased effluent concentration levels.

Sensitivities to Effluent Concentration under Varying Adsorption Rates and Initial Soil P - Turkey Creek Watershed The impact of OWS on stream and soil water P concentration was not significant, as demonstrated by scenarios above. Sensitivity analysis showed that P concentration in Turkey Creek was not sensitive to septic tank effluent discharge concentration at the calibrated values of soil adsorption isotherm and initial soil P.

Further analysis explored how sensitivities to septic effluent discharge could be affected if adsorption and initial soil P values were altered from the calibrated values. The two parameters altered were the most relevant, namely, soil adsorption isotherm and the initial concentration of phosphorus in the soil. The stream P concentration was most sensitive to septic effluent discharge especially when soil adsorption is changed followed by adsorption to suspended sediment. The effect of initial soil P was not as high as for soil adsorption. The results showed that P concentration in stream was sensitive to septic tank effluent if the soil P adsorption is reduced. Initial correlation analysis also showed that the septic effluent discharge has a relatively higher correlation with soil and suspended sediment adsorption isotherms. This could be the reason for P output to be sensitive to the septic effluent discharge when adsorption values are lowered.

11.2.2.3 Nitrate Related Scenarios

The principal form of nitrogen found in ground and surface water is nitrate. In most watersheds, agriculture is very much at the center of concern for nitrate pollution. Septic systems are another anthropogenic source of nitrogen contamination. Ammonium is the major form of nitrogen discharge from septic tank effluent; however, it is eventually converted to nitrate in the soil. In recent years, the potential for groundwater and surface water pollution from individual onsite wastewater disposal systems has emerged as a serious concern in the United States.

OWS are considered one of the major causes for groundwater pollution. Groundwater contamination with nitrogen from OWS may occur due to poor purification of the effluent as a result of insufficient biochemical and physical processes, such as denitrification, and ammonium adsorption. Nutrients released from septic systems into groundwater could be discharged into surface waters. Ammonium and organic nitrogen are the major constitutes leaving the septic tank. Otis et. al. (1973) indicated that the nitrogen in effluent leaving septic tanks is about 75% ammonium nitrogen and 25% organic nitrogen. According to Wilhelm et al. (1994), nitrate-nitrogen causes most of the primary impacts from onsite systems. Because of its mobility, it leaches through aerobic soil profiles to the water table and into the groundwater. In the absence of denitrification, nitrate can flow with the groundwater into adjacent surface waters that serve as groundwater discharge zones and can result in nitrogen contamination of surface waters from onsite systems (Buetow, 2002).

A watershed model can calculate the non-point source loads of OWS that reach the surface waters. It can also predict water quality of the receiving waters due to the combined effect of all point and non-point source loads, including those contributed by OWS. The pollutants discharged undergo treatment in the soil via adsorption and chemical and biological reactions. Watershed scale modeling that accounts for each of these sources and processes can serve as a useful tool for tracking the fate of OWS pollutants discharged through soil to the receiving waters of a river basin.

This study evaluated the impact of different scenarios related to OWS on nitrate concentration both in stream and in soil water using WARMF model. These scenarios are discussed below.

Impact of Population Growth - Turkey Creek Watershed This scenario evaluated the impact of residential development and increase in population using septic tanks on stream nitrate concentration. The population of the TCW, which is currently about 11,000, and the area under residential houses, was increased. Two population level scenarios were evaluated; scenario A, a 50% increase in population, (16,500 people) and scenario B, a 100% increase in population (22,000 people).

The percent change in nitrate concentration as a result of the change in population level from 11,000 to 16,500 and 22,000 was computed. For scenario A, the increase in average nitrate concentration for the two stream segments, R61 and R13, was 11.3 and 13.3 % respectively. The increase for scenario B, for R61 and R13, was 22.1 and 25.9% respectively.

The soil water nitrate concentration was evaluated at catchment C13. The increase in soil water nitrate concentration for Catchment C13, for the bottom soil layer for scenario A and B was 12.84 and 25.7%, respectively. The results for nitrate concentration for the base scenario and for the two population scenarios and the percent change in nitrate concentration are shown in Table 11-4.

The results showed that OWS impacts on nitrate concentration are generally higher than impacts on phosphorus.

Impact of Population Growth - Upper Deschutes Watershed The same population growth scenarios were evaluated for UDW. Nitrate concentration was evaluated at the stream segment near Harper's bridge and at the river mile 13.7 (R12). Nitrate concentration was also evaluated for soil water for catchment 12 (C12).

The results show that there is a slight increase in average nitrate concentration for the two stream segments. There was a 1.42 and 2.8% increase and a 1.05 and 2.07% increase for 50 and 100% population increase, respectively. The change in total P concentration was zero for the same scenario indicating that septic systems have more impact on stream nitrate concentration than phosphorus. However, the % change in nitrate concentration is still much lower than the 12 and 23% increases modeled in the TCW. The results are summarized in Table 11-4.

The lower impact in the UDW, especially on stream concentration, could be attributed to the higher flow rate and the resulting dilution of OWS effluent by the Deschutes River. However, the impacts on stream concentration are generally higher than impacts on phosphorus for both watersheds. The soil water nitrate concentration for C12 for the bottom soil layer increased by 11% and 21% respectively for an increase in population of 50% and 100%.

Table 11-4. Effect of Population Growth on Nitrate Concentration.

Watershed	Location	Scenario	Total P concentration (mg/l)	% change
TCW	Stream Segment-R61	Base Scenario	0.3970	--
		Scenario A	0.4417	11.26
		Scenario B	0.4849	22.12
	Stream Segment R13	Base Scenario	0.2869	--
		Scenario A	0.3250	13.26
		Scenario B	0.3614	25.96
	Bottom soil layer (C13)	Base Scenario	1.3310	--
		Scenario A	1.5020	12.84
		Scenario B	1.6740	25.77
UDW	Stream Segment-Harper's Bridge	Base Scenario	0.0194	--
		Scenario A	0.0197	1.42
		Scenario B	0.0200	2.81
	Stream Segment R12	Base Scenario	0.0208	--
		Scenario A	0.0210	1.05
		Scenario B	0.0212	2.07
	Bottom soil layer (C12)	Base Scenario	0.4440	--
		Scenario A	0.4930	11.00
		Scenario B	0.5380	21.00

Impact of OWS Effluent Concentration - Turkey Creek Watershed This scenario involved increased septic effluent discharge concentration for the existing or base population level (11,000). An effluent concentration of 58 mg/l was used as representative of a standard septic system (base condition), and a new scenario with effluent concentration of 178 mg/l (the 100 percentile value) were evaluated. As shown in Table 11-5, the average percent increase in stream nitrate concentration as a result of increased effluent concentration level from 598 mg/l to 178 mg/l was 33.47% and 35.99%, respectively, for the two stream segments R61 and R13. The soil water nitrate concentration for catchment 13 (C13) for the bottom soil layer increased by 37.5%.

Impact of OWS Effluent Concentration - Upper Deschutes Watershed The same scenario was evaluated for UDW. Nitrate concentration was evaluated at the stream segment near Harper's bridge and at the river mile 13.7 (R12) and for soil water at catchment 12 (C12). The results for nitrate concentration for the base scenario and for new effluent concentration scenario are shown in Table 11-5. The results show that the increase in nitrate concentrations is lower than the results from TCW; however, it is considerably higher than the increase in phosphorus concentration.

Table 11-5. Effect of Septic Tank Effluent Concentration.

Watershed	Location	Scenario	Percentile	Effluent conc. (mg/l)	Nitrate conc. (mg/l)	% change
TCW	Stream Segment-R61	Base scenario	50	58	0.3970	--
		Scenario 2	100	178	0.5259	32.47
	Stream Segment R13	Base scenario	50	58	0.2869	--
		Scenario 2	100	178	0.3902	35.99
	Bottom soil layer (C13)	Base scenario	50	58	1.3310	--
		Scenario 2	100	178	1.8300	37.59
UDW	Stream Segment-Harper's Bridge	Base scenario	50	58	0.0194	--
		Scenario 4	100	178	0.0208	7.49
	Stream Segment R12	Base scenario	50	58	0.0207	--
		Scenario 4	100	178	0.0224	8.24
	Bottom soil layer (C12)	Base scenario	50	58	0.4930	--
		Scenario 4	100	178	0.5810	17.86

Impact of Conversion of OWS to Conventional Sewers -Turkey Creek Watershed The model was used to evaluate the effect of conversion of OWS to sewer systems. Two scenarios were evaluated. Scenario A involved conversion of 50% of the population using septic systems to sewers and scenario B involved a 100% conversion to sewers.

For scenario A, the increase in average nitrate concentration for the two stream segments, R61 and R13, was 13.44 and 5.70%, respectively. The increase for scenario B, for R61 and R13, was 26.40 and 11.48%, respectively.

The soil water nitrate concentration was evaluated at catchment C13. The soil water nitrate concentration for Catchment C13, for the bottom soil layer decreased by 14.80 and 26.40% for scenario A and B, respectively. The results are summarized in Table 11-6.

Impact of Conversion of OWS to Conventional Sewers - Upper Deschutes Watershed Nitrate concentration was evaluated for the two stream segments (stream segment at Harper's bridge and stream segment R12) and soil water for catchment C12. The results show that the percent change in average nitrate concentration for the two stream segments was much higher than the percent change for phosphorus. The results are summarized in Table 11-6.

The increase in average stream nitrate concentration as a result of conversion was 4.01% and 3.09% for stream segment at Harper's bridge and stream segment R12, respectively, when 50% of the population was converted to sewer use. The increase was 11.3 and 9.13% for segment at Harper's bridge and stream segment R12, respectively, when 100% of the population was converted to sewer use. This percent increase in nitrate concentration is higher than the results for total P, which is less than 1% for both scenarios and for both stream segments. There was a reduction in the soil water concentration for bottom soil layer as a result of conversion to sewers. Just like phosphorus, conversion of OWS to sewers may increase stream nitrate concentration and decrease soil water concentrations.

Table 11-6. Effect of Conversion of OWS to Sewers.

Watershed	Location	Scenario	Nitrate concentration (mg/l)	% change
TCW	Stream Segment-R61	Base Scenario	0.3970	--
		Scenario A	0.4504	13.44
		Scenario B	0.5018	26.40
	Stream Segment-R13	Base Scenario	0.2869	--
		Scenario A	0.3033	5.70
		Scenario B	0.3199	11.48
	Bottom soil layer (C13)	Base Scenario	1.3310	--
		Scenario A	1.1340	-14.80
		Scenario B	0.9210	-30.80
UDW	Stream Segment-Harper's Bridge	Base Scenario	0.0194	--
		Scenario A	0.0202	4.01
		Scenario B	0.0216	11.3
	Stream Segment R12	Base Scenario	0.0208	--
		Scenario A	0.0214	3.09
		Scenario B	0.0226	9.13
	Bottom soil layer (C12)	Base Scenario	0.4443	--
		Scenario A	0.4343	-2.25
		Scenario B	0.3493	-21.39

Impact of Soil Nitrification Rates - Turkey Creek Watershed The average stream concentration at different soil nitrification rates was evaluated for the TCW. There is a relatively high correlation between the effect of septic effluent ammonium discharge and other soil and crop related parameters, namely, the nitrification rate, the base saturation for ammonia, cation exchange capacity, initial concentration of ammonium in the soil, leaf composition of ammonium, litter fall rate and the crop productivity parameter, in that order. The correlation implies that the effect of septic effluent ammonium discharge on stream nitrate concentration is dependent on the values used for these parameters.

The effect of septic tank effluent discharge was evaluated at different levels of nitrification (the parameter which was found to have the highest correlation or with the highest influence on the effect of the septic effluent ammonium concentration). The results show that the effect of septic tank effluent ammonium discharge increases with increasing nitrification rate.

Sensitivity analysis of the TCW model showed that the resulting nitrate concentrations in Turkey Creek were not sensitive to the denitrification rate, which indicates that conditions in the TCW do not favor denitrification.

Thus, nitrification rate and cation exchange capacity of the soil were the most important parameters controlling stream and soil water nitrate concentration. The results show that the effect of septic tank effluent ammonium concentration increases with increasing nitrification rate, indicating that unless there are conditions or "hot spots" in the watershed for denitrification, nitrification can only increase stream and soil water concentrations.

Impact of OWS on Soil Water Nitrate Concentration - Turkey Creek Watershed The effect of the same scenarios discussed above for the TCW (population growth and performance of septic tanks) was also evaluated for soil water nitrate concentrations using the WARMF model. WARMF generates concentrations at different soil layers leaving the catchment, which may differ from what is within the catchment, but which can also be an indicator of the concentration in groundwater.

Water quality data during the flow calibration period, September 1998 through November 1999 (Bossong et al., 2003), shows that median concentrations of water nitrite plus nitrate in groundwater (1.0 mg/l) were higher than those of surface water (0.5 mg/l). Lower nitrate plus nitrite concentrations in surface water were attributed to seasonal biological consumption. The mean nitrate concentrations in groundwater reported was 2.2 mg/l. Nitrate concentrations from WARMF model calibrated for surface water nitrate concentration in the bottom of the soil layer, for flow leaving the third soil layer was 1.3 mg/l. Soil water nitrate concentration for the bottom soil layer increased by 6.8 and 13.8% respectively for an increase in population by 50% and 100%.

APPENDIX A

REVIEW OF SCREENING AND GIS-BASED MODELS FOR WATERSHED-SCALE ASSESSMENTS OF ONSITE WASTEWATER SYSTEM IMPACTS

A.1 Introduction

A screening model is any model that evaluates a system under highly simplified relationships between the system components. In the case of a watershed model, the system components include nutrient inputs, watershed characteristics, human population/activity, and nutrient outputs. When completing watershed scale modeling of onsite wastewater systems (OWS), screening models can identify potentially sensitive areas, highlight areas where more data are needed, or quickly test alternative conceptual models. Screening models that apply simplifying and conservative assumptions (e.g., no mixing or attenuation) can evaluate the potential for OWS or other nutrient sources to impact the receiving waters. In other words, a screening model allows the modeler to understand worst-case scenarios within the range of model certainty. This type of model can determine whether more resources should be devoted to data gathering or toward development of more complex models or quantitative tools. Consequently, screening models provide cost-effective evaluations on strategies before more expensive physically-based models are employed for a specific site.

The nutrient load received at the outlet of a watershed will depend on:

- ***Nutrient inputs*** These are sources of nutrients within the watershed. There are four major sources of nutrients in a watershed: atmospheric deposition, fertilizers, wastewater, and fixation of nitrogen by vegetation. The relative contributions of these sources to nutrient loading at the watershed-scale will vary depending on both watershed characteristics and human population/activity within the watershed.
- ***Watershed characteristics*** Topography, geologic material, soil texture/composition, vegetation, and climatic variables will affect the transport, fate, and transformation of nutrients within the watershed.
- ***Human population/activity*** Human population, number of single-family homes, number of individuals per single-family home, water use per person, and land use characteristics are important factors for estimating nutrient loading.
- ***Nutrient outputs*** Watershed processes that store or remove nutrients from the hydrologic system must be accounted for in watershed models.

Nitrogen and phosphorus are the major nutrients of concern in OWS effluent, and consequently are the focuses of watershed-scale modeling efforts. Phosphorus is less of a concern because most OWS users have switched to phosphorus-free detergents, and phosphorus is normally fixed by the soil particles; thus, reducing transport of phosphorus through the subsurface hydrologic system. Phosphorus that is transported to surface water is generally bound to sediment, or predominantly transported in overland flow and runoff rather than groundwater.

Recent studies on OWS nutrient loading to waterbodies have therefore been focused on nitrogen, especially on the usual end product: nitrate.

Screening models differ substantially in the terms and processes included (Valiela et al., 2002) because they were designed for various purposes. For example, some models were designed to estimate P loads from agriculture, while others were designed to estimate N loads from OWS. However, because nutrient loads may be derived from a variety of sources and from both surface water and groundwater flow, the following models have been generalized so that they may be applicable to all cases. The screening models presented below would likely be implemented without rigorous calibration against measured nutrient concentrations/loads.

Examples include Nutrient Loading Model (NLM), Method for Assessment Nutrient-loading And Geographic Evaluation of non-point pollution (MANAGE) and Pollution Load Screening Model (PLSM).

- The NLM, developed by Valiela et al. (1997), calculates nutrient load (mass/time) based on a mass balance approach that accounts for atmospheric deposition, fertilizer use, degradation and plant uptake. The components used in mass balance can be related to land use, population density, and other human activities. Many of the parameters are expressed in units that include a spatial component (e.g., per hectare), so a GIS may be required to perform spatial calculations.
- The MANAGE (Kellogg et al., 1996) estimates total dissolved nitrogen loadings from surface runoff and nitrate from groundwater. This model has two major components designed to deal separately with N loads transported via groundwater and via runoff. The MANAGE model is computed in a spreadsheet, but requires the use of a GIS to derive many of the input variables. A majority of the input variables are related to acreage of land use and fractions of nutrient inputs that runoff from the various land use. GIS-based land use data are available for all parts of the conterminous U.S., while each of the nutrient loading and runoff variables could be estimated from literature values in the absence of watershed specific measurements.
- The PLSM computes annual pollutant loads as a function of runoff volume and mean pollutant concentrations (Adamus and Bergman, 1995). The runoff is calculated from precipitation, and the runoff coefficients vary with land use. The model combines rasters of the land use and soil hydrologic groups to generate a runoff coefficient (RC) based on an input lookup table (Table A.1). The nutrient concentration in runoff also depends on land use. The PLSM uses a lookup table (Table A.2) for mean concentrations from various land use categories based on Adamus and Bergman, (1995).

The MANAGE, PLSM and other GIS-based models require precipitation data to compute runoff. The NCDC cooperative station data can provide the precipitation data.

When implementing screening models, it is often necessary to use reasonable parameter values from the literature, such as those shown in Table A.1 and A.2. However, models that use literature data are not appropriate for prediction. Calibration or an uncertainty analysis should be conducted if a model is used for prediction.

A.2 Types of Models

A.2.1 Mass-Balance Screening Models

Mass-balance screening models usually produce a conservative estimate of nutrient loads, which allows the user to either be: 1) satisfied that there is not a problem (due to conservative estimate) and end the modeling process; or 2) unsure of severity of problem so continue refining the "answer" by implementing more complex model types. Mass-balance screening models are normally implemented in a spreadsheet and are highly simplified estimates of water and nutrient balances. Mass-balance screening models are built upon data that are available as tabular information and may not contain specific geographical locations. Average annual values are most appropriate for mass-balance screening models because data are most often available on this time interval. Each data input variable should be representative of the entire watershed.

A.2.2 GIS-Assisted Screening Models

GIS-assisted screening models are built from a combination of tabular information and geographically-referenced data (i.e., GIS data). The GIS-assisted screening model resembles the mass-balance screening model, but may require the use of spatial analyses available in GIS. Some input variables may be derived from GIS data and then included in the computations of the basic screening model. This type of model allows for more complex modeling and the use of more site-specific data.

A.2.3 GIS-Based Screening Models

GIS-based screening models combine a simple conceptual model and analytical equations that can be solved within a GIS. These types of models are implemented completely within a GIS by processing geographically referenced nutrient inputs, watershed characteristics, human population/activity, and nutrient outputs. A GIS-based screening model generally captures the regional spatial characteristics of a system without including the detailed processes or data density that would be necessary for numerical or physically-based models. An advantage of the GIS-based screening model is that the inputs and outputs can be represented geographically and visualized to aid the modeler in analysis and refinement of the model.

A.3 Model Descriptions and Formulations

A.3.1 Mass-Balance Screening Models

A.3.1.1 Population and Nutrient Level Regression (P&NLR) Model

A simple regression model of nutrient levels (measured at the outlet of the watershed) versus population can be developed when temporal data for surface water discharge, nutrient concentration, and population are available. Model formulation is:

$$NL = \frac{Q * C}{P} \qquad \text{(A.3.1.1-1)}$$

where: NL = nutrient load expressed in units of mass per time-person ($MT^{-1}person^{-1}$),

Q = surface water discharge expressed in units of volume per time (L^3T^{-1}),

C = nutrient concentration expressed in units of mass per volume (ML^{-3}), and P = population expressed in units of persons (Cole et al., 1993). With the regression model, nutrient concentrations may be estimated at points in time by replacing P with previous populations or projected populations.

A.3.1.2 Onsite and Fertilizer (OSF) Model

The On-Site and Fertilizer (OSF) model was applied for land management purposes in Martha's Vineyard, Massachusetts (Gaines, 1986) and is the simplest screening model. OSF assumes that N inputs in the watershed are from homes, and simply adds an annual load of septic N (6.8 kg N a^{-1}) and lawn fertilizer N (4.8 kg N a^{-1}) per household. Atmospheric N loads are assumed to be taken up within the watershed (Valiela et al., 2002). The number of homes in the watershed is the only data requirement, so it will provide a worst-case scenario for N loading to the mouth of the watershed. This model was developed to provide first-order estimates of total dissolved nitrogen (TDN) to receiving waters (Gaines, 1986). Inputs include onsite septic systems and household fertilizer, so the model assumes that human population and households are the primary sources of nutrients (e.g., agricultural fertilizer and atmospheric deposition are insignificant inputs). Model formulation is:

$$NL = H * (SL + FL) \qquad \text{(A.3.1.2-1)}$$

where: NL = nutrient load expressed in units of mass of nutrient per time (MT^{-1}),

H = number of houses in watershed,

SL = the septic load expressed in units of mass of nutrient per time per house ($MT^{-1}H^{-1}$), and FL is the fertilizer load expressed in units of mass of nutrient per time per house ($MT^{-1}H^{-1}$).

A.3.1.3 Horizontal Plane Source (HPS) Model

The Horizontal Plane Source (HPS) model is a transient, three-dimensional analytical model, capable of simulating a horizontal-dispersive movement in a homogenous, isotropic aquifer (Heatwole and McCray, 2007). This model is less complex and requires fewer data inputs than other existing modeling approaches that have been proposed, making it a suitable model for local and county environmental agencies in evaluating OWS groundwater impacts at a development scale.

A.3.2 GIS-asssisted Screening Models

A.3.2.1 Nutrient Loading (NL) Model

The Nitrogen Loading (NL) model (Valiela et al., 1997; Valiela et al., 2002) predicts total dissolved N loads to shallow estuaries from rural suburban watersheds where groundwater flow is the dominant transport vehicle. The NL model is complex enough to represent the nature of the systems and sufficiently adapted to local conditions to produce accurate predictions, while simple enough to be applied to different types of situations. NL model uses values for per capita contributions to estimate N inputs from DWTS that are derived from GIS data representing the total number of residences. NL model also accounts for losses of nitrogen within septic systems (by denitrification, volatilization of ammonia, or by adsorption of ammonium) as well as within leaching fields.

This model was developed for nitrogen loading to estuaries (Valiela et al., 1997) but may be modified for other nutrients or receiving waters where input and loss rates can be estimated. NL model first estimates nutrient input by atmospheric deposition, fertilizer use, and wastewater to major land uses (e.g., natural vegetation, turf, agricultural land, residential areas, and impervious surfaces). Losses are then estimated in various compartments of the watershed (e.g., vegetation, septic systems, soil, vadose zone, and aquifer) with differing loss rates dependent upon land use. Generalized model formulation is:

$$NL = AD + FA - UP - DEG \quad \text{(A.3.2.1-1)}$$

where: NL = nutrient load expressed in units of mass of nutrient per time (MT^{-1}),

AD = atmospheric deposition expressed in units of mass of nutrient per time (MT^{-1}),

FA = fertilizer applied expressed in units of mass of nutrient per time (MT^{-1}),

UP = plant uptake expressed in units of mass of nutrient per time (MT^{-1}), and DEG = degradation expressed in units of mass of nutrient per time (MT^{-1}). Each of the terms on the right-hand side of the generalized model formulation consists of a number of parameters related to land use, population density, and other human activities for a total of 44 input variables. Many of the parameters are expressed in units that include a spatial component (e.g., per hectare), so a GIS may be required to perform spatial calculations.

A.3.2.2 Method for Assessment, Nutrient-loading, and Geographic Evaluation (MANAGE) of Non-Point Pollution Model

The Method for Assessment, Nutrient-loading, and Geographic Evaluation (MANAGE) of non-point pollution model (Kellogg et al., 1996) is a spreadsheet (i.e., Microsoft Excel) based model that uses input derived from spatial analysis in GIS. This uncoupled (Excel to GIS) watershed assessment tool can evaluate pollution risks of land use and landscape features.

MANAGE functions include the identification of areas where natural features and high intensity land uses together increase the risk of nutrient runoff to aquifers and surface waters. It compares the effects of existing and future land use patterns on water resources, and evaluates the effectiveness of storm and wastewater management practices for reducing pollution risk.

The MANAGE model consists of two components for assessing nitrogen contributions to groundwater from OWS, surface water and groundwater. The surface water component of MANAGE uses published export coefficients to estimate N and P loads from 21 land use types. The groundwater component assumes that 80% of the N in OWS enters the aquifer without estimating any losses of NO_3.

MANAGE was developed to estimate total dissolved nitrogen loadings from surface runoff and nitrate from groundwater (Kellogg et al., 1996). This model has two major components designed to deal separately with N loads transported via groundwater and via runoff.

The MANAGE model is computed in a spreadsheet, but requires the use of a GIS to derive many of the input variables. A majority of the 53 (total) input variables are related to acreage of land use classifications and fractions of nutrient inputs that runoff the various land use classifications. GIS-based land use data are available for all parts of the conterminous U.S., while each of the nutrient loading and runoff variables could be estimated from literature values in the absence of watershed specific measurements.

A.3.3 GIS-Based Screening Models

A.3.3.1 Pollution Load Screening (PLS) Model

The St. Johns River Water Management District created the Pollutant Load Screening (PLS) model to cope with Florida's ever increasing population and the resulting pressure on the water quality of lakes and rivers. The PLS model is a GIS-based watershed model for estimating runoff and annual pollutant loads (Adamus and Bergman, 1995). Runoff for each land use and soils combination within the study area is first determined by multiplying average annual rainfall, a runoff coefficient that depends on soil and land use type, and the area of the basin under study. The annual pollutant load is then determined by multiplying runoff by a runoff pollutant concentration coefficient that depends on the type of land use.

This model was developed to estimate annual pollutant loads as a function of runoff volume and mean pollutant concentrations (Adamus and Bergman, 1995). The pollutant load is computed by combining raster-based GIS layers of land use, soil hydrologic groups, and annual precipitation with lookup tables for runoff coefficients and pollutant concentrations. The model was derived for use in Florida, and though it has been revised several times may not be suitable for other climates or conditions. The first step in the PLS modeling process is to combine rasters of the land use and soil hydrologic groups to generate a runoff coefficient (RC) raster, based on the combinations stored in a lookup table (Table A-1):

Table A-1. Runoff Coefficients (RC) Per Land Use/Soil Combination (Adamus and Bergman, 1995).

	Soil Hydrologic Group			
Land Use Category	A	B	C	D
Low Density Residential	0.25	0.30	0.35	0.40
Medium Density Residential	0.30	0.37	0.43	0.50
High Density Residential	0.50	0.57	0.63	0.70
Low Intensity Commercial	0.60	0.70	0.80	0.90
High Intensity Commercial	0.65	0.75	0.85	0.95
Industrial	0.60	0.70	0.80	0.90
Agriculture (all types)	0.15	0.23	0.32	0.40
Mining	0.20	0.30	0.40	0.50
Recreation, Open Space, Range	0.10	0.17	0.23	0.30

The remaining model formulation is:

$$RO = PR * AC * RC$$

where RO is runoff expressed in units of L^3, PR is precipitation expressed in units of L, AC is the area of the cell expressed in units of L^2, and RC is runoff coefficient as a percentage of the precipitation on the cell.

$$NL = RO * PC \qquad (A.3.3.1\text{-}1)$$

where: NL = nutrient load expressed in units of M, RO is runoff expressed in units of L^3 and PC = pollutant concentration expressed in units of ML^{-3}. PC is based on mean concentrations from various land use categories and is multiplied by the RO raster using a lookup table (Table A-2).

Table A-2. Mean Pollutant Concentrations (PC) Expected in Runoff (Adamus and Bergman, 1995).

	Pollutant (mg/L)					
Land Use Category	TN	TP	BOD	SS	Zn	Pb
Low Density Residential	1.77	0.18	4.40	19.1	0.032	0.058
Medium Density Residential	2.29	0.30	7.40	27.0	0.057	0.091
High Density Residential	2.22	0.47	10.60	71.0	0.055	0.091
Low Intensity Commercial	1.18	0.15	8.20	81.0	0.111	0.158
High Intensity Commercial	2.83	0.43	17.20	94.3	0.170	0.214
Industrial	1.79	0.31	9.60	93.9	0.122	0.202
Pasture	2.48	0.48	3.83	55.3	0.028	0.025
Crops	2.68	0.42	3.83	55.3	0.028	0.025
Citrus	2.05	0.14	3.83	55.3	0.028	0.025
Agriculture - Other	2.32	0.34	3.83	55.3	0.028	0.025
Mining	1.18	0.15	9.60	93.9	0.122	0.202
Recreation, Open Space, Range	1.25	0.05	1.45	11.1	0.006	0.025

The PLS model could be used for various pollutants/nutrients, but would ideally be adjusted to site specific pollutant concentrations and include site specific land use classifications. Under the current formulation, the model does not directly account for septic systems, but indirectly accounts for them by having higher TN and TP pollutant concentrations in highly concentrated residential areas.

A.3.3.2 Land Use Nutrient Loading (LUNL) Model

The model presented here is a modified version of a phosphorus loading model designed to account for spatial pattern of land use.

A.4 Summary

Several of the models described above, in addition to others, can be implemented using a web-based modeling tool (http://nload.mbl.edu) hosted by the Marine Biological Laboratory in Woods Hole, Massachusetts. The modeling tools available on this site (summarized in Table A-3) are designed for modeling of nitrogen loading to estuaries so may not be perfectly suitable for modeling loading of other nutrients.

Table A-3. Summary of Modeling Tools Available on the NLOAD Website.

Model	Screening Model Type	Reference	Number of Input Variables	N result	Form of N predicted
OSF	Mass Balance	Gaines, 1986	3	N load	TDN
CC	Mass Balance	Caraco and Cole, 1999	11	N load	NO_3
PJM	GIS-assisted	Johnes, 1996	32	N load	TDN
NLM	GIS-assisted	Valiela et al., 1997	44	N load	TDN
MANAGE	GIS-assisted	Kellogg et al., 1996	53	N load	NO_3
BBP	GIS-assisted	Costa et al., 1999	41	N load	TDN
CPCP	Mass Balance	Cole et al., 1993	3	N concentration	DIN
DVM	Mass Balance	Dettmann, 2001	4	N concentration	DIN
ELM	Mass Balance	Valiela et al., 2004	32	N concentration	DIN

When implementing screening models, it is often necessary to use reasonable parameter values from the literature. The values shown in Table A-2 could be used as input variables when watershed specific data are unavailable.

Various models have been developed to study transfer of nutrients from the land surface and subsurface to receiving waters. The approaches range from simple (regression models that relate nutrient loads to watershed characteristics) to complex (physically-based models that account for all of the known processes for simulating the water cycle and nutrient cycle based on climate, topography, soil properties, land use, and management practices).

A screening model can estimate nutrient concentrations or loading to receiving waters, assess relative contributions from a variety of nutrient sources, and evaluate sensitivity of nutrient concentrations to various contributing variables. Screening models can evaluate large watersheds before implementing more detailed, time consuming and data intensive modeling approaches (Adamus and Bergman, 1995). Screening models are the simplest in design and require a reasonable number of input parameters. A screening model captures major watershed characteristics without including detailed processes that would be necessary for site-specific delineations. When integrated with GIS, screening models become powerful and time-efficient since, with the power of the storage, manipulation, analysis, and visualization of geographically-referenced data, a GIS can handle site-specific problems.

APPENDIX B

REVIEW OF DISTRIBUTED MODELS USED FOR WATERSHED-SCALE ASSESSMENTS OF ONSITE WASTEWATER SYSTEM IMPACTS

B.1 Introduction

Watershed models have only sparsely and recently been used for assessing watershed-scale problems associated with Decentralized Wastewater Treatment Systems (DWTS). Efforts to use watershed models to compare the impact of DWTS with other sources of point and non-point pollution are very limited. The development of many models has placed a new burden on model users: selecting the appropriate model.

The selection of the right model under certain constraints requires a comprehensive knowledge of the capabilities and features of available models. As non-point source pollution has garnered more attention in recent years, governmental agencies, academic and research institutions, and consulting firms have developed methods of assessing pollution from non-point sources. Many of these methods involved the modification and development of hydrology and water-quality models. Recently, some of these models have been linked with geographic information systems (GIS) for ease of data management. A literature review identified potential and appropriate models for watershed-scale DWTS problems.

An investigation of the recent models that handle point and non-point source pollution is included. This review evaluates and summarizes some of the key features of the most widely cited watershed-scale, hydrodynamic and water quality models with the emphasis on DWTS. Model selection was based on minimum criteria. Emphasis was given to models that can simulate the most common wastewater pollutants, especially nitrogen and phosphates both at a field and watershed scale. Sediment transport was also considered since transport and fate of sediments and nutrients are intimately related. The models reviewed are AGNPS, ANNAGNPS, ANSWERS-2000, CREAMS-WT, GLEAMS, HSPF, MIKE-SHE, SWAT, MODFLOW, SWMM, WARMF, WMS, and GIS screening models.

B.2 Review of Watershed Models

This section describes some of the most commonly used non-point source pollution models. For each model, a general description of the model is discussed. Also provided are overview model requirements, outputs, model capabilities and limitations, procedures used to simulate sediment, nutrients and pesticides, runoff, subsurface-flow and groundwater flow interactions, model suitability for an onsite wastewater systems, model history of application, model availability, and model simplicity for users. Tables B-1 and B-2 summarize many important model features. Finally, a summary is provided on all models with focus on relevance to DWTS.

B.2.1 AGNPS (Agricultural Non-Point Source Pollution Model)

B.2.1.1 Description

AGNPS simulates runoff, sediment and nutrient transport from watersheds that have agriculture as their prime use. AGNPS is a distributed parameter, event-based model (Young et al., 1995) that operates on a grid cell basis. AGNPS is developed by the U.S. Department of Agriculture to evaluate the effect of management decisions in agricultural watershed-scale systems including potential impacts of point and non-point source pollution on surface water quality. To implement model computations, a watershed is subdivided into a grid of square elements. Each element, typically about 100 m^2, requires 22 parameters to describe its antecedent conditions, physical characteristics (e.g., soil properties and slope steepness), management practices and rainfall.

B.2.1.2 Model Type

AGNPS is a distributed model in the sense that watershed geometry is represented by discrete elements (uniformly distributed squares). AGNPS is an event model. It computes the response of the catchment to the storm event, including loadings in the runoff. It does not track changes in the watershed between storm events nor does it estimate annual reductions. Although the basic structure of the model is process-based, the key components are empirical; hence it is classified as a statistical model. Model inputs and outputs are listed in Table B-1.

B.2.1.3 Model Capabilities and Limitations

Spatial variation: In AGNPS the watershed is divided into square cells. Subdivision of main cells into smaller sub-cells gives flexibility to account for heterogeneity in the watershed. Due to this discretization, all watershed characteristics are expressed at the grid-cell level thus requiring the input of spatially distributed data that is handled through the use of Geographical Information Systems (GIS) (León, L.F. and Lam, D.C. 2000).

Temporal variation: AGNPS is an event model. It is initialized with soil and landscape conditions prior to occurrence of a storm and computes the response of the catchment to the storm event, including loadings in the runoff. It does not track changes in the watershed between storm events (León, L.F. and Lam, D.C. 2000).

Model Limitations: The model considers surface hydrology, stream flow and infiltration, but sub-surface hydrology is ignored. This can be a serious limitation with sandy soils and high water table soils. The model does not allow the input of spatially variable rainfall data. This can be a limitation as the size of the watershed increases (León et al., 2004). Storm event precipitation is considered uniform throughout the watershed.

Hydrology and water quality outputs: AGNPS can simulate surface runoff, sediment loading, nitrogen, phosphorus and pesticides.

B.2.1.4 Procedures Used to Compute Model Essential Outputs

Runoff: Runoff is calculated with the standard SCS curve number method.

Sediment: Upland erosion and sediment transport is estimated using a modified form of the Universal Soil Loss Equation, USLE (Wischmeier and Smith, 1978). Surface sediment mobilization is computed from the USLE and transport capacity is based upon Bagnold stream power equation. Sediment is routed from cell to cell through the watershed to the outlet using a sediment transport and depositional relationship described by Foster et al. (1981), which is based on a steady-state continuity equation.

Nitrogen: AGNPS, SWAT, CREAMS, GLEAMS and HSPF have similar routines for nutrient transformation. Nitrogen, phosphorus, and pesticides in these models are based on routines developed for the CREAMS/GLEAMS models including biochemical processes and groundwater loading. Transformation and transport of nutrients and pesticides is presented in more detail under SWAT model in the subsequent sections and a brief discussion given under each model.

The two main nutrient types simulated in AGNPS are nitrogen and phosphorus. Nutrient transformation in both cases is governed by nutrient cycle. A daily mass balance of inputs by fertilizer or residue decomposition and loss by plant uptake, transport or transformation for each element is computed for each cell. Major components considered are uptake of N by plants, application of fertilizers, residue decomposition, and soil N transformations. The day's sediment-bound N and soluble N in runoff are determined for the cell. Nitrogen is partitioned into organic and mineral parts and a separate mass balance computed for each. N cycles are simplifications that track only major N transformations of mineralization from humidified soil organic matter and plant residues, crop residue decay, and fertilizer and plant uptake. The transformation of N in the soil profile is adapted from EPIC model. Plant uptake of N is modeled through a simple crop growth stage index. Chemical transport of N in both dissolved and suspended (adsorbed) forms from the watershed by runoff.

Phosphorus: P transformation is governed by phosphorus cycle. Plant residue and fertilizer are the source of phosphorus in the soil. Organic phosphorus from crop residue is mineralized to soluble phosphorus, which is available for uptake, by plants. Like nitrogen transport of P is occurs in both dissolved and adsorbed forms by runoff.

Pesticides and other organic contaminants: A daily mass balance adapted from GLEAMS (Leonard et al., 1987) is computed for each pesticide. AGNPS allows simulations for any number of pesticides, each with its own independent chemical properties. Each pesticide is treated separately; independent equilibration is assumed for each pesticide. Major components of the pesticide model include foliage wash-off, vertical transport in the soil profile, and degradation. Soluble and sediment-adsorbed fractions are calculated for each cell on a daily basis (see SWAT model pesticide simulation below for more information).

B.2.1.5 Model Availability and Simplicity for Users

Model is available in public domain. In its original form, much labor was necessary to develop the input file(s) for AGNPS, but in recent years some work has been carried out coupling AGNPS to a GIS framework. ArcView Geographic Information System (GIS) can be used for data extraction, and processing of critical area. Geospatial databases built with geographic information systems (GIS) serve as primary sources of input data to AGNPS. Elevation, land cover, and soil data provide the input parameters required by the AGNPS are extracted. Several researchers have integrated AGNPS and Geographic information Systems (Mitchell et al., 1989; Panuska et al., 1991; He et al., 1993; Srinivasan and Arnold, 1994).

B.2.1.6 History of Application

AGNPS has been widely used in recent years, especially in the Midwest. McIntosh et al. (1993) applied three models (EPIC, AGNPS and SWRRB) to evaluating sediment-loading reductions in dairy farm watersheds in Wisconsin. Holmberg et al. (1998) made comparative runs of AGNPS and EPIC to evaluate fertilizer application strategies. Sugiharto et al. (1994) describe a similar exercise in applying AGNPS and EPIC to sediment and phosphorus loadings under twenty different management strategies for dairy farm dominated watershed. Foerster and Milne-Home (1995) report an application in Wales to a "conservation tillage trial site." Corbett et al. (1997) evaluated AGNPS for two watersheds beyond the types for which the model was initially developed for a forested watershed in coastal South Carolina and an urbanized watershed, examining storm water runoff volumes, flow rates, and sediment loads.

B.2.1.7 Runoff, Subsurface Flows, and Groundwater Flow

The model focuses on predictions of surface water flow and sediment yield for single precipitation events only (Grunwald and Norto, 1999). The model considers only surface water, and subsurface flow is ignored. There are no mass balance calculations tracking inflow and outflow of water. This can be a serious limitation with sandy soils, high water table soils, or soils with other unfavorable characteristics. There is no groundwater routine in AGNPS, hence baseflow contributions to stream flow is ignored.

B.2.1.8 Model Suitability for Onsite Wastewater Systems

There is no direct of way of introducing septic system effluent in AGNPS. In addition, AGNPS doesn't have groundwater routine. As mentioned earlier, the model considers surface hydrology, stream flow and infiltration, but sub-surface hydrology is ignored. AGNPS has a nitrogen and phosphorus transformation and transport component that considers the chemical process such as mineralization, nitrification, denitrification and plant uptake and leaching and also the relevant processes for phosphorus (uptake, sorption etc). However, because Onsite Wastewater System (OWS) effluent is transmitted via subsurface, the model is not generally suitable to simulate treatment and transformation of OWS pollutants. The model assumes that percolation below the soil profile would be lost from the system and ignored. AGNPS does not compute subsurface or groundwater flow or concentrations of chemicals in groundwater. Furthermore, for OWS pollutants most important effect on stream would be during base flow but AGNPS is only an event model.

B.2.2 ANNAGNPS (Annualized Agricultural Non-Point Source Pollution Model)

B.2.2.1 Description

AnnAGNPS model was later developed as a continuous simulation model using the event based AGNPS as a basis. Another modification made on AGNPS model was incorporation of subsurface lateral flow, including a subsurface drainage feature. The remaining features are the same as the AGNPS.

B.2.2.2 Model Type

Like AGNPS, it is a distributed model but unlike AGNPS (which is a model useful for analyzing and evaluating storm event processes) it a continuous model and can be used for analyzing long-term effects of hydrological changes and watershed management practices, especially agricultural practices. Model inputs and outputs are similar to AGNPS.

B.2.2.3 Model Capabilities and Limitations

Spatial variation: Like AGNPS the watershed is divided into square cells. All watershed characteristics are expressed at the grid-cell level thus requiring the input of spatially distributed data that is handled through the use of Geographical Information Systems.

Temporal variation: AnnAGNPS is a continuous simulation model. It can be used for analyzing long-term effects of hydrological changes and watershed management practices.

Model Limitations: AnnAGNPS doesn't have groundwater routines. It doesn't calculate concentrations in groundwater.

Hydrology and water quality outputs: AnnGNPS can simulate surface runoff, subsurface flow, sediment loading, nitrogen, phosphorus and pesticides.

B.2.2.4 Procedures Used to Compute Model Essential Outputs

Procedures used to compute runoff, sediment, nutrients and pesticide loadings are similar to AGNPS.

B.2.2.5 Model Availability and Simplicity for Users

Model is available in public domain. Some work has been carried out coupling AnnAGNPS to a GIS framework.

B.2.2.6 History of Application

Yuan et al. (2001) tested the Deep Hollow watershed. The test was used to validate the runoff and sediment yield predictions of the AnnAGNPS pollutant loading model and to evaluate the effectiveness of Best Management Practices (BMPs) in terms of sediment reduction. Over a three-year period, predicted runoff was 89% of actual runoff and model sediment yield was 104% of observed.

Yuan et al. (2003) again tested the AnnAGNPS Version 2.0 suite of models on the Deep Hollow Watershed, this time focusing on the nitrogen loading component. After the simulation, a sensitivity analysis was performed, showing that the initial N concentration in the soil and the N uptake rate by the crop (cotton) had the greatest impact on N loading from the watershed. Comparisons between observed and predicted N loading rates from the watershed revealed that AnnAGNPS performed poorly in simulating N loading.

B.2.2.7 Runoff, Subsurface Flows and Groundwater Flow

Previously AnnAGNPS also simulated surface runoff with percolation as the only subsurface flow process, with no subsurface lateral flow, subsurface drainage, or base flow. The contribution of lateral flow and ground flow to stream runoff was ignored. It had been assumed within AnnAGNPS that percolation below the soil profile would be lost from the system and ignored. However, these flows can be significant in areas with soils having high hydraulic conductivities in surface layers and a water-restricting layer below.

Subsequently, subsurface lateral flow, including a subsurface drainage feature, was incorporated into a new version of AnnAGNPS. Subsurface lateral flow was defined based on Darcy's equation. Subsurface drainage was determined using Hooghoudt's equation, which describes subsurface flow as a function of saturated hydraulic conductivity, distance between drains, water table height above the drains and equivalent depth of the impermeable layer below the drain. Subsurface lateral flow and subsurface drainage were assumed to occur only when the soil becomes saturated.

Soil moisture content within AnnAGNPS is based on the water balance equation, which incorporates a simple bookkeeping of inputs and outputs during a day. The water input includes snowmelt, precipitation, and irrigation water. For the second soil layer, the water input is the percolation from the first layer that occurs if the soil moisture content in a layer exceeds field capacity. The hydraulic gradient is approximated by the local surface topographic slope. The soil profile is assumed as isotropic and the saturated lateral hydraulic conductivity is the same as the saturated vertical hydraulic conductivity.

B.2.2.8 Model Suitability for Onsite Wastewater Systems

AnnGNPS was developed from the AGNPS model. As mentioned above, the modifications include converting the model to a continuous incorporation of subsurface lateral flow, including a subsurface drainage feature. In additions to this, AnnAGNPS simulates transformations of nutrients using the same process described for AGNPS. There is no direct of way of introducing septic system effluent in AnnAGNPS. However, because AGNPS can simulate subsurface flow, it could simulate OWS better than AGNPS. Users can input the amount of nitrogen and phosphorus that is actually applied in the cell. Hence, it might be possible to assume that nitrogen released from septic tanks to the septic system drain fields is converted to nitrates in the soil and make an estimate of the effect of onsite septic systems. AnnGNPS also has a provision for inputting point fertilizer source from point sources such as feedlots.

B.2.3 ANSWERS-2000 (Areal Non-Point Source Watershed Environment Response Simulation)

B.2.3.1 Description

Beasley et al. (1980) developed the original ANSWERS as a new version of ANSWERS. ANSWERS-2000 can evaluate the effects of land use, management schemes and conservation practices or structures on the quantity and quality of water from both agricultural and non-agricultural watersheds. It was initially developed on a storm event basis. The current version of the model, ANSWERS-2000, is a continuous simulation model that was developed in the mid 1990s (Bouraoui and Dillaha, 1996). In this version, the nutrient sub models were overhauled and improved infiltration routine (Green and Ampt) was incorporated and soil moisture and plant growth components were added to enhance long-term continuous simulation.

B.2.3.2 Model Type

ANSWERS-2000 is a long-term, continuous simulation, physically based, distributed parameter watershed model. A continuous model can be used for analyzing long-term effects of hydrological changes and watershed management practices compared to an event used for analyzing processes during storm events. Model inputs and outputs are presented in Table B-1.

B.2.3.3 Model Capability, Strength and Limitations

Spatial variation: ANSWERS has a distributed structure that allows handling spatial as well as temporal variability of pollution sources and loads.

Temporal variation: It was initially developed on a storm event basis intended to apply to runoff processes during and immediately after a rainfall event. The current version (ANSWERS-2000) is a long-term, continuous simulation model.

Model Limitations: The model cannot be applied in large watersheds and longer simulations. Because ANSWERS-2000 is primarily for surface water, it doesn't work equally well for all land uses and soil types.

*Hydrology and water quality outputs:*ANSWERS-2000 can simulate surface runoff, sediment loading, nitrogen, and phosphorus.

B.2.3.4 Procedures or Routines Used to Compute Model Essential Outputs

Runoff: The Green-Ampt infiltration equation forms the basis of infiltration component in the ANSWERS-2000 model. The model does not address the dynamics of water in the subsurface zone particularly well. Specifically, there is no saturated soil overland flow, and flows from the subsurface contributing to channel output are assumed a linear function of the storage. The model accounts for percolation out of the single soil layer that is used to model the profile, assuming a simplified linear relationship between percolation and storage.

Sediment: Sediment transport in ANSWERS-2000 differs from the one used in AGNPS (modified soil loss equation). ANSWERS-2000 model was revised by Bouraoui and Dillaha (1996) to include channel scour. Furthermore, the subroutines governing rill and inter-rill erosion and transport were altered to include critical-shear subroutines. The modules are adapted from the WEPP model, and are considered sufficiently process based. Channels are modeled in much the same way as rills, with critical shear stress theory (Dabral and Cohen, 2001). The module for computing the capacity of overland flow to carry sediment is retained from the former ANSWERS model. Sediment transport is based on the equations of Yalin (1963).

Nitrogen: ANSWERS-2000 models nitrogen dynamics. The formulation includes are re-active organic N, ammonium and nitrate. Nitrogen bound in plant tissue and in plant residue and stable organic nitrogen are also included, and sources and sinks of N include fertilizer and atmospheric N. Nitrogen mineralization (organic N to nitrate) is modeled as a two-step process, via storage of ammonium.

Phosphorus: Phosphorus is modeled using five pools. They are re-active organic P, labile P, active mineral P, passive mineral P, and fresh organic P (Bouraoui and Dillaha, 1996).

Pesticides and other organic contaminants: one of the important limitations is the fact that no pesticides are considered. So an important step recommended in the improvement of the ANSWERS model would be the introduction of a pesticide component following the procedures used to simulate the nutrients (Bouraoui and Dillaha, 1996).

B.2.3.5 Model Availability and Simplicity for Users

Model is available in public domain. A user-friendly interface to ANSWERS-2000 called QUESTIONS runs on the Windows platform using Visual Basic 6.0, ArcView GIS 3.2, and Map Objects 2.0. By leading the user through a series of simple forms and GIS layers. QUESTIONS constructs and formats detailed ASCII files used by ANSWERS-2000 for each simulation run.

This interface improves the likelihood that the ANSWERS-2000 input file correctly reflects the user desired variable values (Dabral and Cohen, 2001).

B.2.3.6 History of Application

ANSWERS-2000, was tested on two watersheds in Watkinsville, Georgia, and performed well in predicting runoff, sediment, nitrate, dissolved ammonium, sediment-bound TKN, and dissolved phosphorus losses from both watersheds (Dillaha et al., 2001). The model did not predict sediment-bound ammonium losses from either watershed well. The model was also tested on the 1153 ha Owl Run watershed in Virginia. It performed well for the largest storms, and cumulative predictions of runoff volume, sediment yield, nitrate, ammonium, sediment-bound TKN, and orthophosphorus were within 40% of the measured values.

B.2.3.7 Runoff, Subsurface Flow and Groundwater Flow

ANSWERS 2000 is primarily a runoff model. As described earlier, the model does not address the dynamics of water in the subsurface zone particularly well. Because the original model was event-oriented, the focus has been on carefully simulating what happens when it rains, and little attention has been paid to the interim periods, except insofar as they provide the antecedent conditions for the next rainfall event. The model, therefore, is considered inappropriate for high-base flow conditions. There is a coefficient in the model for dealing with tile drainage systems. This provides the model a constant rate of daily tile drain flow for all conditions.

B.2.3.8 Model Suitability for Onsite Wastewater Systems

Just like AGNPS, ANSWERS-2000 is primarily a runoff model. As stated earlier, the model does not have good routines to address the dynamics of water in the subsurface zone and percolation. It doesn't have routines for groundwater flow and concentrations in groundwater. So contributions of chemicals to surface flow by lateral flow and groundwater flow are not considered. ANSWERS-2000 doesn't have a procedure to introduce septic system effluent. It also poorly represents alternative fertilizer application methods. It can't simulate point sources inputs such as animal waste application or management effects. Hence, this model is not suitable for simulating OWS pollutants.

B.2.4 CREAMS-WT

B.2.4.1 Description

The CREAMS-WT model can simulate pollutant movement on and from a field site, including such constituents N and P, pesticides and sediment (Knisel, 1980). The effects of various agricultural practices can be assessed by simulation of the potential water, soil, nutrient and pesticide losses in runoff from agricultural fields. The spatial scale of the model is the size of an agricultural field. The model structure consists of three major components: hydrology, sedimentation and chemistry. The hydrology component estimates the volume and rate of runoff, evapotranspiration, soil moisture content and percolation (Heatwole et al., 1987).

B.2.4.2 Model Type

CREAMS-WT is a physically based model, lumped parameter, and continuous simulation model. Model inputs and outputs are presented in Table B-1.

B.2.4.3 Model Capability and Limitations

Spatial variation: CREAMS-WT is a field scale model for predicting runoff, erosion, and chemical transport from agricultural management systems. It is applicable to field-sized areas. A field is defined in the context of the CREAMS model as a management unit having a single land use, relatively homogeneous soils, spatially uniform rainfall, and single management practices.

Temporal variation: CREAMS-WT is a continuous simulation model that uses a daily time step for infiltration and soil water movement and shorter time step for storms.

Model Limitations: Model application is limited to field size area. Because of its field-scale focus, CREAMS-WT is limited to representing only surface runoff contributions to stream flow. Subsurface flow and infiltration losses of chemicals - are not tracked.

Hydrology and water quality outputs: CREAMS-WT can simulate surface runoff, sediment loading, nitrogen, phosphorus and pesticides.

Model Limitations: The model cannot be applied in large watersheds and longer simulations. It doesn't simulate subsurface flow contributions to stream flow.

Hydrology and water quality outputs: CREAMS-WT can simulate surface runoff, sediment loading, nitrogen, and phosphorus.

B.2.4.4 Procedures or Routines Used to Compute Model Essential Outputs

Runoff: The hydrologic component consists of two options. When only daily rainfall data are available, the SCS curve number model is used to estimate runoff. If hourly rainfall data are available, an infiltration-based model is used to simulate runoff. Water movement through the soil is modeled using a simple approach, with flow occurring when a layer exceeds field capacity.

Sediment: The erosion component maintains elements of the USLE, but includes sediment transport capacity for overland flow.

Nitrogen: The plant nutrient routine has a nitrogen component that considers mineralization, nitrification, and denitrification processes. Plant uptake is estimated, and nitrate leached by percolation out of the root zone is calculated. Furthermore, the nutrient component use enrichment ratios to estimate the portion of the nitrogen transported with sediment.

Phosphorus: Components of phosphorus cycle such as mineralization and plant uptake are considered. Enrichment ratio is used to estimate the portion of phosphorus transported with sediment.

Pesticides or other organic contaminants: The pesticide component considers foliar interception, degradation, and wash off, as well as adsorption, desorption, and degradation in the soil. Several of the equations developed for the CREAMS model have been used or modified within other models (Skaggs, 1997).

B.2.4.5 Model Availability and Simplicity for Users

The model is available in public domain capable of copying and distribution. Spatially distributed data can also be input to CREAMS-WT.

B.2.4.6 History of Application

Many applications in the literature use the model as the basic model for landscape processes in an agricultural setting, reporting comparative studies, or employing the CREAMS output in a project of larger scope. Applications include comparing conservation BMP strategies, both nonstructural (no-till, contour, contour with waterways, strip crop with waterways, filter strips) and structural (terraces, tile outlet terraces, sediment basins).

B.2.4.7 Runoff, Subsurface Flow and Groundwater Flow

The model uses a soil storage routing technique to predict the flow through the root zone. It divides the root zone into layers for routing. CREAMS-WT predicts runoff, evapotranspiration, percolation and soil water content. Soil water movement and seepage are simulated. The movement of the water table in the soil profile is very critical to follow, because the absence of any such track on the water movement will result under-prediction or over-prediction of water storage in soil, resulting in the under or over-prediction of runoff. Because of its field-scale focus, CREAMS-WT is limited to representing only surface runoff contributions; subsurface and infiltration of chemicals are simply removed from the computational system.

B.2.4.8 Model Suitability for Onsite Wastewater Systems

CREAMS-WT has routine for nutrient transformation and transport similar to AGNPS. Creams-WT doesn't have a procedure to account for contributions from OWS pollutants. Because of its field-scale focus, CREAMS is limited to representing only surface runoff contributions; subsurface and leaching losses are predicted but do not contribute to stream flows because the losses are simply removed from the computational system and do not contribute to the groundwater base flow loading. CREAMS-WT does not compute subsurface or groundwater flow contributions to stream flow. It doesn't calculate chemical contributions from subsurface or groundwater to stream flow. The CREAMS-WT model can simulate pollutant movement on and from a field site, including such constituents as fertilizers (N and P), pesticides and sediment in surface runoff, however due to the limitations mentioned above the model is not considered suitable for simulating OWS pollutants

B.2.5 GLEAMS (Groundwater Loading of Agricultural Management Systems)

B.2.5.1 Description

GLEAMS was developed to simulate edge-of-field and bottom-of-root-zone loadings of water, sediment, pesticides and plant nutrients from complex climate-soil-management interactions. The model is an extension of the CREAMS model with an enhanced hydrology component, which includes the vertical flux of pesticides, and a comprehensive plant nutrient model. Like CREAMS-WT, GLEAMS is a field-scale model where a field is defined as a management unit having a single land use, homogeneous soil, spatially uniform rainfall and a single management practice (Muller and Gregory, 2003). The model simulates vertical flux of soluble tracers into the subsurface, with much greater detail in the soil percolation process than CREAMS. Details of the hydrology of the model are given by Knisel and Williams (1995). While GLEAMS does not include the vadose zone per se, it was designed with the intent to be coupled to an appropriate vadose zone model. One of the main purposes for the development of GLEAMS was to better simulate the leaching of pesticides from agricultural operations (Knisel and Williams, 1995; Cohen, 1996).

B.2.5.1 Model Type

GLEAMS is a physically based, lumped parameter, continuous simulation model. GLEAMS inputs are similar to CREAMS. Inputs and outputs are presented in Table B-1.

B.2.5.2 Model Capability and Limitations

Spatial variation: GLEAMS is a field-scale model for predicting runoff, erosion, and chemical transport from agricultural management systems. It is applicable to field-sized areas. GLEAMS assumes that a field has homogeneous land use, soils, and precipitation.

Temporal variation: GLEAMS is a continuous simulation, which operates on daily basis.

Model Limitations: Model application is limited to field size area. The model has better routines to track movement of pesticides especially with percolated water. However it doesn't calculate the contribution of the subsurface and groundwater flow to chemical loading to the streams.

Hydrology and water quality outputs: GLEAMS can simulate surface runoff, sediment loading, nitrogen, phosphorus and pesticides in stream flow and in groundwater.

B.2.5.3 Procedures or Routines Used to Compute Model Essential Outputs

Runoff: The hydrology component simulates runoff due to daily rainfall using a modification of the SCS curve number method. Hydrologic computations for evapotranspiration, percolation, infiltration, and runoff are determined using a daily time step (Leonard et al., 1987).

Sediment: The erosion component is similar to the one developed for the CREAMS model (Knisel, 1980). This component considers overland, channel, impoundment, or any combination of these routes. The model uses the universal soil loss equation (USLE) and the concept of continuity of mass to predict erosion and sediment transport under different topographic conditions.

Nitrogen: GLEAMS nutrient component includes the processes and transformations, considering surface and subsurface pathways to estimate their losses from the field. The user can select among different management alternatives, such as land application of animal wastes in solid, slurry, or liquid form through irrigation as well as in applied commercial fertilizers. Besides this, GLEAMS can simulate pesticide application, its degradation in the foliage and soil, the extraction into runoff and the transport with sediments, also the losses due to evaporation and plant uptake. The nutrient component of the GLEAMS model is a complex sub model and considers both nitrogen and phosphorus cycles. The nitrogen component includes: mineralization, immobilization, denitrification, ammonia volatilization, nitrogen fixation by legumes, crop N uptake, and losses of N in runoff, sediment, and percolation below the root zone. It also considers fertilizer and animal waste application.

Phosphorus: Similarly, the phosphorus component includes: mineralization, immobilization, crop uptake, losses to surface runoff, sediment and leaching, and it also includes fertilizer and animal waste application.

Pesticides and other organic contaminants: The pesticide component of the GLEAMS model is designed to allow simulation of interactions among pesticide properties, soils, climate, and management and the effects on pesticide losses in surface runoff, attached to transported sediment, and in percolation below the root zone or any other specified depth. To trace the fate of surface applied or incorporated pesticides, GLEAMS considers degradation, adsorption, and convective processes in each of the computational soil layers in the root zone. Upward movement of pesticides due to evaporation and plant uptake is also included. Initial pesticide residue in the soil is an input to the model. The movement of the pesticide is controlled by its solubility, degradation half-life, and soil organic carbon adsorption coefficient.

Characteristics of pesticide adsorption to soil organic carbon are used to partition compounds between solution and soil fractions for simulating extraction into runoff, sediment, and percolation losses. Pesticide dissipation in soil and on foliage is treated as a first-order process with a different apparent half-life for each.

B.2.5.4 Model Availability and Simplicity for Users

GLEAMS is available in public domain and it is capable o copying and distribution. There are several strategies to link a model to a GIS, going from loose coupling to a total integration, through close coupling (Tim U.S., 1996; Corwin, 1996).

B.2.5.5 History of Application

GLEAMS has been used extensively in different parts of United States. Craig and Weiss (1993) used GLEAMS to simulate pesticides entering surface water from USDA Forest Service nurseries - to the stream water. Goss (1992) used GLEAMS to evaluate various categories and combinations of pesticide loss. Leonard et al. (1992) applied GLEAMS to evaluate potential pesticide runoff of two similar pesticides from one soil, for the purpose of comparing annual means and single events. Neary et al. (1993) report on ten years of watershed-scale research on pesticides in forested watersheds throughout the southern U.S. They used data on various forestry pesticides to verify GLEAMS, CREAMS, and PRZM models. Shirmohammadi and Knisel (1994) validated GLEAMS against leaching data from lysimeter experiments conducted in soil near Uppsala, Sweden. They conclude that the GLEAMS model performed "in a reasonable manner." Shirmohammadi et al. (1998) reported extended work on nutrient losses through tile drainage with similar results. Zacharias and Heatwole (1999) evaluated the pesticide-prediction performance of GLEAMS and PRZM using field data from a plot under no-till corn in the Coastal Plain region of Virginia. Yoon et al. (1994) applied GLEAMS to predict nutrient (N and P) losses in surface and subsurface runoff, and their concentrations in soil layers on corn plots at the Tennessee Valley Substation in Alabama. The GLEAMS simulation was compared to field data and it was found that both soluble and sorbed P losses in surface runoff and NO3-N in leachate and soil layers "were not consistent with field data." The predicted N losses were too high, and the predicted P concentrations in leachate were too low.

B.2.5.6 Runoff, Subsurface Flow and Groundwater Flow

Daily soil-water accounting is done each day to obtain the soil-water content and, thus, the available storage, on any day of rainfall, snowmelt, and/or irrigation. Water redistribution in the soil is calculated by computational soil layer using soil water retention, and water transmission characteristics. Infiltration of rainfall and irrigation is not explicitly calculated in GLEAMS, but is implicit in the application of the SCS curve number procedure. Water movement through each computational layer and percolate out of the bottom of the root zone are calculated using a storage-routing technique. Travel time through each layer is estimated from layer thickness and saturated conductivity. Time of percolation (drainage of excess water to field capacity) is estimated from the volume of percolate and travel time through the layer. Volume of percolation is not determined by saturated conductivity, except under restricted drainage (low saturated conductivity) when water in excess of field capacity cannot drain out of a layer in one day. This results in a simulated high water table, reduced percolation, and increased evapotranspiration. The lack of free and unrestricted drainage may result in reduced solute transport as well.

B.2.5.7 Model Suitability for Onsite Wastewater Systems

GLEAMS has routines to compute transformation of nitrogen and phosphorus. GLEAMS has been developed with special emphasis on vadose zone processes to represent movement of chemicals to groundwater. Water movement through each computational layer and percolate out of the bottom of the root zone are calculated using a storage-routing technique. The model tracks movement of pesticides with percolated water, runoff, and sediment. The focus of the model is surface runoff and leaching of water and agricultural chemicals rather than on base flow contributions. Just like the models discussed so far, there is no direct of way of introducing septic system effluent in GLEAMS but the model is relatively suitable to evaluate the effect of OWS pollutants on groundwater through leaching using the fertilizer application routine to apply equivalent amount of nitrates and phosphates that may be applied through OWS pollutants. The model could be used to estimate the loading pollutants to the groundwater table. However, if the objective is to determine the effect of OWS pollutants on the stream water quality, the model is not suitable since subsurface and baseflow contribution are not simulated.

B.2.6 HSPF (Hydrological Simulation Program - FORTRAN)

B.2.6.1 Description

Johansen et al. (1984) developed the Hydrological Simulation Program - FORTRAN (HSPF) model. A continuous-time model, HSPF allows simulation of contaminant runoff, stream water quality and sediment interaction formulations. HSPF is a comprehensive model that allows the integrated simulation of land and soil contaminant runoff processes within stream hydraulic and sediment-chemical interactions (Donigian and Huber, 1991). HSPF is based upon the concepts of the mechanistic Stanford Watershed Model, and originally incorporated aspects of the early watershed models ARM (Agricultural Runoff Management) and NPS (Non-Point Source) (Donigian et al., 1995). Overall subdivision of the watershed into computational catchments is based upon distribution of meteorological stations and soil types, which are considered to define segment groups, each of which is assumed homogeneous in climatologic and soils. Each such group is further subdivided according to land use classifications, which can include vegetation, agricultural cropping patterns and urbanization. Another purpose of this modular design was to allow HSPF to be readily coupled to water-quality models. The total load from HSPF includes the contribution from groundwater and overland flow.

B.2.6.2 Model Type

HSPF is continuous simulation, physically based, distributed parameter watershed model. Major input requirement and outputs from the model are listed in Table B-1.

B.2.6.3 Model Capability and Limitations

Spatial variation: HSPF is a distributed, physically based model that combines spatially distributed physical attributes into hydrologic response units (HRUs), each of which, in response to meteorological inputs such as precipitation, potential evapotranspiration, and temperature and storage-capacity factors such as interception, surface retention, and soil-moisture storage, is assumed to behave in an uniform manner.

Temporal variation: HSPF is a continuous watershed model that simulates processes for extended periods of time just like other continuous simulation models. It simulates hydrologic, and associated water quality processes on pervious and impervious land surfaces and in streams and well-mixed impoundments

Model limitations: HSPF works better for in-stream processes, but weaker for upland (overland) processes compared to other models like SWAT. It has a groundwater transport routine but weaker compared to MIKE SHE and MODFLOW.

Hydrology and water quality outputs: HSPF can simulate surface runoff, sediment loading, nitrogen, phosphorus and pesticides.

B.2.6.4 Procedures or Routines Used to Compute Model Essential Outputs

Runoff: HSPF is an infiltration-excess model that separates moisture inputs (precipitation and snowmelt) into infiltrating and non-infiltrating fractions. Overland flow is simulated by the Chezy-Manning equation and average values of the surface roughness, length, and slope for the overland flow plane of each Hydrologic Response Unit (HRU). Subsurface lateral flow has a substantial effect on storm flow hydrographs, particularly in areas where vertical percolation is retarded by bedrock or a shallow or poorly permeable soil layer. Subsurface lateral flow is termed interflow–outflow in HSPF. The partitioning between these two pathways (infiltration and direct runoff) is a function of the soil moisture and the infiltration rate. Potential direct runoff is further partitioned as either direct surface runoff, interflow runoff, or upper zone storage. Runoff volume, peak flow, infiltration, evapotranspiration, soil water content, and percolation are computed on a daily basis. If detailed precipitation data are available then infiltration is calculated at histogram breakpoints.

Sediment: HSPF models sediment production and removal for pervious land areas. The processes include detachment, wash-off of the detached sediment, and scour of the soil matrix. Detachment is modeled as a power function of rainfall rate and wash-off of the detached sediment, and scour of the soil matrix are modeled as a power function overland flow rate.

Nitrogen: Plant nutrients and pesticides are simulated and storm load and average concentrations of sediment-associated and dissolved chemicals are determined in the runoff, sediment, and percolation through the root zone (Leonard and Knisel, 1984). HSPF simulates soil nitrogen transformations. This includes plant uptake of nitrate and ammonium, return of plant nitrogen to organic nitrogen, denitrification, immobilization of nitrate-nitrite and ammonium, mineralization of organic nitrogen, fixation of atmospheric nitrogen, volatilization of ammonium, adsorption/desorption of ammonium, and partitioning of two types of organic nitrogen between solution and particulate forms. Adsorbed ammonium and particulate forms of organic nitrogen are removed from the surface layer storage by association with sediment. Dissolved nitrate, ammonium, and dissolved organic nitrogen are transported with water.

Phosphorus: The method used to transport and react phosphorus is the similar to that used for nitrogen. HSPF simulates the behavior of phosphorus in a pervious land segment. This involves modeling the transport, plant uptake, adsorption/desorption, immobilization, and mineralization of the various forms of phosphorus. Because phosphorus is readily tied to soil and sediment, it is usually scarce in streams and lakes. The method used to transport phosphorus is the same as that used for nitrogen. Organic phosphorus and adsorbed phosphate are removed with sediment and phosphate in solution is transported in the moving water.

Pesticide and other organic contaminants: HSPF simulates the processes of degradation and adsorption as well as transport. The pesticides are simulated in the soil and runoff in three forms: dissolved, adsorbed, and crystallized. These phases in the soil affect the forms and amounts in the runoff. Pesticides move with water flow or by association with the sediment. They also may be adsorbed to the soil in varying degree as a function of the chemical characteristics of the toxicant and the exchange capacity of the soil layer. Adsorption and desoprtion are simulated by first-order kinetics, using a single K value Freundlich isotherm or a varying Freundlich K value.

B.2.6.5 Model Availability and Aimplicity for Users

Model is available for copying and distribution in public domain. Model inputs can be generated from commonly available GIS data.

B.2.6.6 History of Application

There have been a number of applications of HSPF. The largest application is the 62,000 square mile tributary area to the Chesapeake Bay. The smallest application has been experimental plots of a few acres near Watkinsville, Ga. The most significant applications have been in the Seattle area, Chicago area, Patuxent River, MD, Truckee-Carson Basins, NV, and watersheds in Pennsylvania.

B.2.6.7 Runoff, Subsurface Flow and Groundwater Flow

As mentioned earlier, HSPF is an infiltration-excess model that separates moisture inputs into infiltrating and non-infiltrating fractions. Subsurface lateral flow (interflow) is also computed. HSPF considers all stream flow components; surface runoff, interflow, and base flow. Runoff volume, peak flow, infiltration, evapotranspiration, soil water content, and percolation are computed on a daily basis. There are three possible fates in the subsurface: interflow, base flow, or root zone storage. The infiltrated water is first partitioned between interflow and active groundwater recharge in a similar way to how surface water is partitioned between surface runoff and infiltration. Active groundwater recharge is further partitioned between base flow storage and root-zone storage and is a function of the index to soil moisture in the model. Root-zone storage terminates in evapotranspiration; base flow storage is routed to the stream through another simple linear reservoir with a longer residence time than for interflow. Inflows to the interflow component may occur from the surface or from upslope external lateral flows.

B.2.6.8 Model Suitability for Onsite Wastewater Systems

HSPF considers all stream flow components; surface runoff, interflow, and base flow. It computes lateral flow and baseflow contributions from groundwater. HSPF simulates concentration of water quality constitutes both in subsurface and groundwater outflow. This makes HSPF a potential model to simulate onsite wastewater systems. Local census data can be incorporated to provide sewered population estimates and calculate potential loading from the watershed as a whole, which may be introduced to the simulated stream network as point sources. Point source loads can be associated with the subbasin in which they are located. However, OWS pollutants cannot be treated as point sources. There is no direct way to simulate OWS pollutant injection, transformation and transport. However, using fertilizer management operation in HSPF, effluent from OWS may be applied as nitrogen fertilizer.

B.2.7 MIKE SHE

B.2.7.1 Description

MIKE-SHE is a modified and expanded version of SHE marketed by the Danish Hydraulic Institute (DHI). MIKE SHE can be used for the analysis, planning and management of a wide range of water resources and environmental problems related to surface water and groundwater. MIKE SHE includes a traditional 2D or 3D finite-difference groundwater model, which is very similar to MODFLOW. MIKE SHE's overland-flow component includes a 2D finite difference diffusive wave approach using the same 2D mesh as the groundwater component. MIKE SHE is used to simulate flow and transport of solutes and sediments in both surface water and groundwater (DHI, 1998). The model consists of several modules depicting the complete terrestrial hydrological cycle: evapotranspiration component, unsaturated zone flow component, saturated zone flow component, overland and channel flow, and irrigation.

B.2.7.2 Model Type

MIKE SHE code is a powerful, physically based, distributed parameter, fully integrated code for three-dimensional simulation of hydrologic systems. Data requirements and outputs are presented in Table B-1.

B.2.7.3 Model Capability, Strength, and Limitations

Spatial variation: MIKE SHE is a fully distributed model. The spatial and temporal variation of meteorological, hydrological, geological and hydrogeological data across the model area is described in girded form for the input as well as the output from the model

Temporal variation: It has been successfully applied at multiple scales, using spatially distributed, and continuous climate data to simulate a broad range of integrated hydrologic, hydraulic and transport problems in humid as well as more arid areas.

Model Limitations: MIKE SHE is relatively complex model and difficult to use. Mike SHE is publicly available for purchase and is relatively expensive. Input and calibration data requirements are very large.

Hydrology and water quality outputs: MIKE SHE can simulate surface runoff, sediment loading, nitrogen, phosphorus and pesticides or other organic contaminants. It simulates overland flow, unsaturated flow, vegetation-based evapotranspiration, and groundwater flow.

B.2.7.4 Procedures or Routines Used to Compute Model Essential Outputs

Flow: Overland flow velocities and water depths are described by the Saint-Venant equation adopting the Manning formulation for friction slope. The model calculates temporal and spatial ponding depths and flows on the ground surface. The MIKE-SHE model computes infiltration in its unsaturated-zone module. MIKE-SHE has three options for simulating the unsaturated zone. The first uses a one-dimensional representation of the governing equation of unsaturated flow in the vertical direction. The second is a simple gravity flow method that ignores capillary suction pressure, and the third method is a simple two-layer water balance model for shallow water tables. The unsaturated flow is linked to the groundwater module to update the location of the water table. It is also linked to the overland flow module to provide water for overland flow when saturation conditions exist.

MIKE SHE includes a traditional 2D or 3D finite-difference groundwater model, which is very similar to MODFLOW. The geology is described in terms of layers or lenses with attached hydraulic properties. Properties can be specified either on a cell-by-cell basis or by property zones. MIKE SHE's overland-flow component includes a 2D finite difference diffusive wave approach using the same 2D mesh as the groundwater component. Overland flow interacts with the river, the unsaturated zone, and saturated groundwater zone. In the river system the temporal and spatial variations in water levels and discharge are produced.

The Mike SHE Erosion Module is an adaptation of the Eurosem model to the MIKE SHE concept. Splash detachment as well as equations of detachment, transport, and sedimentation are included in the soil erosion module. Modeling soil detachment by raindrop impact in EUROSEM is based on relationships between detachment and the kinetic energy of rainfall. Soil detachment by runoff is modeled in terms of a generalized erosion–deposition theory. Sediment concentration in runoff reflects a balance between the two continuous counteracting processes of erosion and deposition. It implies that the ability of flowing water to erode its bed is independent of the amount of material it carries and is only a function of the energy expended by the flow, particularly the shear between the water and the bed, and the turbulent energy in the water. The capacity of runoff to transport-detached soil particles is expressed in EUROSEM in terms of a concentration.

Nitrogen: Nitrogen transformation and transport is simulated in conjunction with DAISY model, developed at the Danish Veterinary and Agricultural University. DAISY produces the output of the unsaturated zone, subsequently transferred as input to the groundwater system in the MIKE SHE. The two models were integrated, so the reactions and temperature calculations take place in DAISY while the flow and solute transport take place in MIKE SHE. DAISY is an advanced soil-plant-atmosphere system column model. It describes crop production as well as water and nutrient dynamics in the root zone of the agro-ecosystems according to various management strategies, including crop rotations, fertilization, irrigation, and soil tillage and crop residue management. The model simulates processes including: plant growth and crop production, heat flux and soil temperature, soil water uptake by plants and evapotranspiration, carbon and nitrogen mineralization, nitrification and denitrification, nitrogen uptake by plants. Combined with the MIKE SHE modules, the DAISY module provides a powerful tool for the assessment of the regional impacts of agricultural crop production system management on water quality conditions in the soil, the groundwater, and streams.

Phosphorus: Phosphorus transformations and transport is also included.

Pesticides and other organic chemicals: With respect to pesticides, another recent development is a further merging between the DAISY process descriptions and pesticide processes. Modeling the transport and metabolism of pesticides requires incorporation of sorption and degradation processes. The MIKE SHE sorption/degradation module includes simplified descriptions of complex geochemical and microbiological processes.

Model Limitations: Limitations are high complexity and difficulty of use. Input and calibration data requirements are very large. The model is also expensive.

Hydrology and water quality outputs. Groundwater and surface water discharge and concentrations can be determined.

B.2.7.5 Model Simplicity for Users

MIKE SHE is relatively difficult to use. But the graphical interface is significantly improved in the 2003 version, offering a dynamic navigation tree, dynamic dialogs, and limited on-line documentation, and notably improved output animation capabilities. It seamlessly links with Arcview shape files and has well-organized spreadsheet-graphical functionality for ease in editing spatial and temporal input. In addition, there is a pre- and post-processing user interface, MIKE SHE PP that includes capabilities such as digitization of mapped contours, river system and areally distributed data, Interpolation routines to provide point values and grid averages, graphical editing of 2-D data and river data, graphical presentation of simulation results in full color graphics, plots of the variations in space of a variable in any layer or along any line through the model, plots of time series of any variable. To take full advantage of MIKE SHE, a user must be an expert in surface water, groundwater and vadose zone models, as well as GIS.

B.2.7.6 History of Application

MIKE SHE has been successfully applied at multiple scales, using spatially distributed, and continuous climate data to simulate a broad range of integrated hydrologic, hydraulic and transport problems in humid as well as more arid areas. In the U.S., it has been used extensively in South Florida on Everglades Restoration projects, in Colorado, at Rocky Flats (a former DOE nuclear manufacturing facility) and in the Black Mesa basin of northeastern Arizona. It has been successfully applied at multiple scales, using spatially distributed, and continuous climate data to simulate a broad range of integrated hydrologic, hydraulic and transport problems in humid as well as more arid areas.

B.2.7.7 Runoff, Subsurface Flow, and Groundwater Flow

MIKE SHE is an integrated surface water/groundwater-modeling tool. MIKE SHE includes process models for overland flow, unsaturated flow, vegetation-based evapotranspiration, groundwater flow, and fully dynamic channel flow. MIKE SHE includes both simple and advanced process models. It has the capacity to explicitly model groundwater flow, overland runoff, ET, wetland flow, river flow, and lake flow. Interaction between surface water and groundwater is well simulated. The model simulates both confined and unconfined aquifers in user specified arrangements using a finite-difference approach. Pumping and leakage between aquifers can also be modeled. The groundwater model is linked dynamically to the surface and unsaturated zone making it suitable for surface-groundwater interaction studies.

B.2.7.8 Model Suitability for Onsite Wastewater Systems

MIKE SHE in combination with other models can be used to describe crop production as well as water and nutrient dynamics in the root zone including crop rotations, fertilization, irrigation, and soil tillage and crop residue management. It can also be used to simulate processes including plant growth, nutrient (nitrogen and phosphorus) transformation processes. MIKE SHE includes routines for overland flow, unsaturated flow, evapotranspiration, groundwater flow, and fully dynamic channel flow. Interaction between surface water and groundwater is well simulated. The groundwater model is linked dynamically to the surface and unsaturated zone making it suitable for surface-groundwater interactions. Hence the model can be a very useful tool to simulate OWS pollutants although it does not have a direct approach to simulate OWS pollutants.

B.2.8 MODFLOW

B.2.8.1 Description

MODFLOW is the general three-dimensional groundwater model developed by the USGS (Hill, 1992). The model currently has little surface-water capability. While surface-water models are soon to be included with the package, these packages do not handle water quality simulations. Therefore the model can serve no immediate role in a TMDL determination. MODFLOW solves the three-dimensional groundwater flow equation for a porous medium by using a finite-difference method. It is currently the most used numerical model for groundwater flow problems. MODFLOW-2000 simulates steady and non-steady flow in an irregularly shaped flow system in which aquifer layers can be confined, unconfined, or a combination of confined and unconfined. Flow from external stresses, such as flow to wells, areal recharge, evapotranspiration, flow to drains, and flow through riverbeds, can be simulated. Hydraulic conductivities or transmisivities for any layer may differ spatially and be anisotropic, and the storage coefficient may be heterogeneous. Pollutant transport and reactions are handled by add-on modules to MODFLOW called MT3D (Zheng, 1993) or RT3D (Clement et al., 1998)

B.2.8.2 Model Type

MODFLOW is a physically based distributed parameter model. Model inputs and outputs are given in Table B-1.

B.2.8.3 Model Capabilities and Limitations

Spatial variation: MODFLOW is a fully distributed model. The spatial and temporal variation of hydrological, geological and hydrogeological data across the model area is described in girded form for the input and output from the model.

Temporal variation: MODFLOW is a continuous simulation model that simulates processes for extended periods of time.

Model Limitations: MODFLOW is relatively complex model with additional modules required for chemical transport.

Hydrology and water quality outputs: MODFLOW can simulate surface runoff, sediment loading, nitrogen, phosphorus and pesticides or other organic contaminants. MODFLOW by itself simulates groundwater flow only.

B.2.8.4 Procedures or Routines Used to Compute Model Essential Outputs

Groundwater flow: The groundwater flow equation is solved using the finite-difference approximation. The flow region is subdivided into blocks in which the medium properties are assumed to be uniform. In plan view, the blocks are made from a grid of mutually perpendicular lines that may be variably spaced. Model layers can have varying thickness. A flow equation is written for each block, called a cell. Several solvers are provided for solving the resulting matrix problem; the user can choose the best solver for the particular problem. Flow-rate and cumulative-volume balances from each type of inflow and outflow are computed for each time step.

Nitrogen: MODEFLOW in conjunction with MT3D can be used to simulate nitrogen transformation and transport. MT3D is a 3D solute transport model for simulation of advection, dispersion, and chemical reactions of dissolved constituents in groundwater systems. The modular structure makes it possible to independently simulate advection, dispersion, sink/source mixing, and chemical reactions without reserving computer memory space for unused options. MT3D is intended for use with any block-centered finite-difference flow model such as MODFLOW. The finite-difference code in MODFLOW can be used to simulate the distribution of hydraulic head within the groundwater. MT3D uses simulated hydraulic heads, intercell flows, and source and (or) sink terms from the MODFLOW output in the solution of the advection-dispersion equation. MT3D can be used to simulate changes in concentration of single-species miscible contaminants in groundwater considering advection, dispersion, and some simple chemical reactions. The chemical reactions included in the model are equilibrium-controlled linear or non-linear sorption and first-order irreversible decay or biodegradation. Nitrate transport was described by the convection–dispersion equation solved using MT3D (Zheng, 1993). The three-dimensional solute-transport model MT3D is used to simulate the concentrations of nitrate. MT3D uses simulated hydraulic heads, intercell flows, and source and (or) sink terms from the MODFLOW output in the solution of the advection-dispersion equation. RT3D is another add-on module that is implemented in a similar fashion as MT3D. An advantage of RT3D is that the user can specify chain reactions or kinetics.

Phosphorus: Dissolved phosphorus in the saturated zone can be modeled using MT3D Pesticides and other organic contaminants: Other models such as RT3D and MT3D can be used in conjunction with MODFLOW to simulate pesticide movement. MT3D allows simulation of sorption with a linear, Freundlich isotherm and simulates degradation assuming pseudo-first order kinetics with a separate degradation rate for sorbed and dissolved phases. Additional reactions can be specified using RT3D as described above.

B.2.8.5 Model Availability and Simplicity for Users

While simple in formulation, MODFLOW is complicated to use because it is designed to include many different modules. However, there are many graphical user interfaces (GUIs) commercially available that are relatively inexpensive. These generally implement MODFLOW and MT3D but usually do not include RT3D on other advance packages. Thus complicated text files must be mastered by the user for these applications. The graphical interface in visual MODFLOW makes it easy to use. Interactive model display in both plan view and cross-sectional view. Some experimental work has been carried out coupling MODFLOW to GIS-based data systems, e.g., Orzol and McGrath (1989) and Tang and Kondoh (1996).

B.2.8.6 History of Application

Bissett and Poeter (1994) describe an application of MODFLOW with the Stream Package to determine the interaction between an aquifer and surface streams near Golden, Colorado. Interaction of groundwater with a wetlands system was simulated using MODFLOW by Bradley and Brown (1997). Swain et al. (1992) describe an application using these new capabilities to modeling an interacting wetlands-river-aquifer system in Dade County, Florida. The model has been coupled with surface-water models for special-purpose applications. Fredericks and Labadie (1993) combined MODFLOW with a river-basin network model MODSIM to evaluate water-volume/flow interactions between surface and groundwater in the South Platte basin. Yan and Smith (1994) coupled MODFLOW with the South Florida Water Management Model (a surface-network model) to simulate surface-groundwater interactions. All of these applications involved surface-ground water volume (or flow) interactions, and none address water-quality. There have been numerous MODFLOW applications in addition to these.

There have been also OWS pollutant related applications. Morgan and Everett (2005) used MODFLOW-MT3D in conjunction with optimization model to estimate the optimal loading of nitrate from decentralized wastewater treatment systems to an aquifer. The method utilizes a simulation-optimization approach in which a nitrate fate and transport simulation model is linked to an optimization model. Using this method, maximum (optimal) sustainable loading rates that meet constraints on groundwater quality and nitrate loading to streams via groundwater discharge was determined. The method was demonstrated in conjunction with the National Onsite Demonstration Project (NODP) in the community of La Pine in southern Deschutes County, Oregon. This research has been done on one of our watershed included in the project for model evaluation and is relevant since it was particularly a modeling research on the impact of OWS.

B.2.8.7 Runoff, Subsurface Flow, and Groundwater Flow

As stated earlier, MODFLOW has no surface-water quality simulation capability and has limitations in its treatment of groundwater/surface-water interaction. The groundwater/surface-water interaction is typically treated very simply. Estimates of net groundwater recharge are applied as a specified-flux boundary condition in the recharge package and an estimated groundwater ET vs. water-table depth function is applied as a boundary condition in the ET Package.

In recent years, more complex groundwater/surface-water simulators such as MIKE-SHE, and HMS-MODFLOW have been developed. These simulators explicitly model groundwater flow, overland runoff, ET, wetland flow, river flow, and lake flow. However, these simulators are significantly more complex than MODFLOW and much more difficult to use. GeoTrans has also developed an open-source MODFLOW called Surface/Vadose Package for MODFLOW, or SV Package module that automatically calculates overland runoff, ET, and net groundwater recharge from given rainfall and anthropogenic inflows (e.g., irrigation, septic-tank discharge). Another model being developed by USGS is GSFLOW, a groundwater/surface-water flow model that couples the Precipitation Runoff Modeling System (PRMS) with MODFLOW. The coupling is via a new soil-moisture zone module in PRMS and a new unsaturated-zone flow package in MODFLOW. Unfortunately, these packages do not handle chemical transport.

B.2.8.8 Model Suitability for Onsite Wastewater Systems

MODFLOW can be used in conjunction with other models such as MT3D or RT3D to simulate changes in concentration contaminants such as nitrates phosphates in groundwater considering advection, dispersion, and some simple chemical reactions. As stated earlier, MODFLOW has no surface-water quality capability and has limitations in its treatment of groundwater/surface-water interaction. Thus this model cannot be used for TMDL application or surface water quality problems. The groundwater/surface-water interaction is typically treated very simple. However, to evaluate the comprehensive effect of OWS pollutants, the movement and transformation of OWS pollutants in the unsaturated vadose zone is also important. This component is missing in MODFLOW. A useful feature of MODFLOW-MT3D is that OWS pollutants can be added as a liquid phase to the subsurface.

B.2.9 SWAT (Soil and Water Assessment Tool)

B.2.9.1 Model Description

SWAT is a river basin, or watershed scale model developed by the USDA-ARS. It is developed to predict the impact of land management practices on water, sediment and agricultural chemical yields in large complex watersheds with varying soils, and land use and management conditions over long periods of time. SWAT operates on daily time step (Arnold et al., 1998). The first level of subdivision in SWAT is the subbasin. A subbasin contains at least one Hydrologic Response Unit (HRU), a tributary channel and a main channel or reach. HRUs are portions of a sub-basin that possess unique land use/management/soil attributes (Neitsch et al., 2002).

B.2.9.2 Model Type

SWAT is a distributed parameter, physically based, continuous simulation model. Model inputs and outputs are presented in Table B-1.

B.2.9.3 Model Capabilities and Limitations

Spatial variation: Subdivision of subwatersheds into smaller HRUs, gives flexibility to account for heterogeneity in the watershed. HRUs are portions of a sub-basin that possess unique land use, management, soil attributes (Neitsch et al., 2002). The possibility of having small and relatively uniform land units (HRUs) can be used to reduce the error due to lumping effects.

Temporal variation: SWAT is a continuous simulation model. The initial conditions for each day are determined by the model based on the conditions on the previous day.

Model Limitations: SWAT is weak in in-stream processes, although it is a good model for upland (overland) processes. SWAT has groundwater transport routine but weaker compared to MIKE SHE and MODFLOW.

Hydrology and water quality outputs: SWAT can simulate surface runoff, lateral flow, groundwater flow, sediment loading, nitrogen, phosphorus and pesticides.

B.2.9.4 Procedures Used to Compute Model Runoff, Sediment, and Nutrients

Runoff: The SWAT hydrology model is based on the water balance equation. A distributed SCS curve number is generated for the computation of overland flow runoff volume, given by the standard SCS runoff equation.

Sediment: Sediment yield is determined from the Modified Universal Soil Loss Equation (MUSLE) (Arnold, 1992). SWAT takes both deposition and degradation into account in sediment routing. Deposition is based on fall velocities of various sediment sizes. Degradation is determined from Bagnold's stream power equation. Sediment size is estimated from the primary particle size distribution (Foster et al., 1980) for soils the SWAT model obtains from the STATSGO (USDA, 1992) database.

Nitrogen: The plant/soil model incorporates nitrification, denitrification, nitrogen fixation, volatilization, mineralization (nitrogen and phosphorus), immobilization (nitrogen and phosphorus), and plant uptake (nitrogen and phosphorus).

Transformation of nitrogen from one form to another is governed by the nitrogen cycle. Once initial levels (specified by the user or initialized by the model) are known, the amount of each form of nitrogen is computed based on nutrient balance and transformations including nitrification, denitrification, immobilization, plant uptake, leaching, and volatilization.

The three major forms of nitrogen in mineral soils are organic nitrogen associated with humus, mineral forms of nitrogen held by soil colloids, and mineral forms of nitrogen in solution. Amounts of NO3-N contained in runoff, lateral flow and percolation are estimated as products of the volume of water and the average concentration of nitrate in the layer. Organic N transport with sediment is calculated with a load function developed by McElroy et al. (1976) and modified by Williams and Hann (1972). The load function estimates the daily organic N runoff loss based on the concentration of organic N in the topsoil layer, the sediment yield and the enrichment ratio. The enrichment ratio is the concentration of organic N in the sediment divided by that in the soil (Neitsch et al., 2002).

Phosphorus: Phosphorus transformation is governed by phosphorus cycle. Plant residue and fertilizer are the source of phosphorus in the soil. The three major forms of phosphorus in soils are organic phosphorus associated with humus, insoluble forms of mineral phosphorus (e.g., precipitated P) and plant-available phosphorus in soil solution. The amount of soluble P and organic phosphorus contained in humic substances for all soil layers can be defined by the user or initialized by the model at the beginning of the simulation. Soluble P removed in runoff is predicted using concentration in the top 10 mm of soil, the runoff volume and a partitioning factor. Phosphorus in soil is mostly associated with the sediment phase. Organic and mineral P attached to soil particles are associated with sediment load from the HRU, specifically to the concentration of phosphorus attached to sediment, sediment yield, and the phosphorus enrichment ratio which is logarithmically related to sediment concentration described by Menzel (1980).

Pesticides and other organic contaminants: SWAT simulates movement of pesticide into the stream via surface runoff in solution and sorbed to sediment, and into the soil profile and aquifer by percolation in solution. The equations used to model the movement of pesticide in the land phase were adopted from GLEAMS (Leonard et al., 1987). Pesticide movement is controlled by its solubility, degradation, and soil organic carbon adsorption coefficient. Pesticide may be aerially applied to an HRU with some fraction intercepted by plant foliage and some fraction reaching the soil or incorporated into the soil through tillage. SWAT models pesticide loss as a result of wash off, degradation, surface runoff, lateral flow and leaching. The majority of pesticides are organic compounds. Pesticide degradation is governed by first order kinetics and depends on the initial amount of pesticide in the soil layer, the rate of degradation and time elapsed since the initial pesticide amount was determined. Water-soluble pesticides can be transported with percolation deep into the soil profile and potentially pollute shallow groundwater systems. Pesticide in the soil can be transported in solution or attached to sediment. The partitioning of a pesticide between the solution and soil phases is defined by the soil adsorption coefficient for the pesticide. Pesticide in the soluble phase may be transported with surface runoff, lateral flow or percolation. The change in the amount of pesticide contained in a soil layer due to transport in solution with flow is a function of time, concentration and amount of flow. Pesticide attached to soil particles may be transported by surface runoff. This phase of pesticide is associated with the sediment yield and changes in sediment loading will be reflected in the loading of sorbed pesticide.

B.2.9.5 Model Availability and Simplicity for Users

Model is available for copying and distribution in public domain. The source is available to users. An ArcView GIS interface is available to generate model inputs from commonly available GIS data. SWAT requires a large number of input files to run. However, data inputting is simplified using GIS. Srinivasan and Arnold (1994) describe integration of SWAT into a GIS system for input data set development and model output visualization. These GIS data are summarized by the interface and converted to a form usable by the model. Elevation, soils, and land use are used to generate the input files. Observed temperature and precipitation are also incorporated.

B.2.9.6 History of Application

Applications of SWAT to several Texas river basins have been carried out by the Blackland Research Center (BRC), including the Lower Colorado, North Concho, and Trinity. SWAT has also been applied in Texas to the watershed of Lake Waco, including the Bosque River basin. The Bosque watershed is affected by numerous diary operations as well as row-crop agriculture and municipal waste discharges. A customized version of SWAT was created by incorporating subroutines and algorithms from APEX (yet another USDA-ARS agricultural water-management model) to better depict soluble P and N fluxes into soil and plants and to improve the modeling of Best Management Practices. A watershed modeling using the SWAT model was performed to understand the potential influence of various point and non-point sources of P in the Blue River watershed (the most developed of three watersheds that supply Lake Dillon). The watershed model was calibrated to measured flow rates and P concentrations. Fertilizer management practice was used to simulate OWS input in SWAT. The mass input rate of OWS pollutants was set equal to the mass of nutrient input by the fertilizer. Because simulations without OWS contributions showed little change in the concentration of P in the stream, OWS are not believed to be the primary source of P in the lake. Instead, P in runoff sediments is the most likely contributor to surface water (Lemonds and McCray, 2003). Pradhan et al. (2005) evaluated fate and transport of nitrogen derived from onsite systems using SWAT model using a similar approach using fertilizer input routine. Geza and McCray (2006) completed analysis using SWAT model on the impact of using the STATSGO soil survey data versus the SSURGO soil survey data for nitrogen, phosphorus and sediment loading using realistic inputs on Turkey Creek Watershed.

B.2.9.7 Runoff, Subsurface Flow, and Groundwater Flow

Soil water content in a layer is calculated based on water balance. Downward flow, or percolation, occurs when field capacity of a soil layer is exceeded and the layer below is not saturated. Lateral flow will be significant in areas with soils having an impermeable layer at a shallow depth. The flow rate is governed by the saturated conductivity of the soil layer.

Lateral subsurface flow, or interflow originates below the surface but above the zone of saturation. SWAT incorporates a kinematic storage model for subsurface flow developed by Sloan et al. (1983). This model simulates subsurface flow in a two-dimensional cross-section along a flow path down a steep hill slope. The model is based on the mass continuity equation, or water mass balance.

SWAT partitions groundwater into two aquifer systems: a shallow, unconfined aquifer which contributes baseflow to streams within the watershed and a deep, confined aquifer which contributes return flow to streams outside the watershed (Arnold et al., 1993). In addition to return flow, water stored in the shallow aquifer may replenish moisture in the soil profile in very dry conditions or be directly removed by plant.

Water enters shallow aquifer storage primarily by infiltration/percolation, although recharge by seepage from surface waterbodies may occur. Water leaves groundwater storage primarily by discharge into rivers or lakes, but it is also possible for water to move upward from the water table into the capillary fringe. The daily water balance for the shallow aquifer depends on the amount of recharge entering the aquifer, the groundwater flow, or base flow into the main channel, the amount of water moving into the soil zone in response to water deficiencies, the amount of water percolating from the shallow aquifer into the deep aquifer and the amount of water removed from the shallow aquifer if there is pumping.

B.2.9.8 Model Suitability for Onsite Wastewater Systems

No algorithms currently exist in SWAT to specifically simulate OWS. However, fertilizer management practice can be used to simulate OWS. Only the fraction of fertilizer applied on the top 10mm interacts with surface water. The remainder is applied below the surface. The fraction can be varied from 0 to1. It might be possible to assume that nitrogen released from septic tanks to drain fields is converted to nitrates in the soil and make an estimate of the effect of onsite septic systems. There are different forms of nitrogen fertilizers in SWAT database ranging from elemental nitrogen to chemical forms of nitrogen such as ammonium nitrate. A fertilizer type that may have similar constituents of OWS effluent can be chosen from SWAT database. Hence, using fertilizer management operation in SWAT, effluent from onsite systems may be applied as nitrogen fertilizer. There is no direct way of injecting liquid into the subsurface. However, SWAT has a provision for applying irrigation water as management operation. This can be useful to simulate OWS effluent. The liquid required for infiltration should be either from precipitation or irrigation because it is not possible to inject liquid waste as non-point pollution source. Thus, while an appropriate amount mass can be introduced to the subsurface, the timing and liquid volume cannot be accurately assessed. In addition, any irrigation water that runs off will artificially contribute to stream flow.

If a sewer is used instead of decentralized OWS the discharge from wastewater treatment source can be considered as a point source discharge into the streams. SWAT directly models the loading of water, sediment and nutrients from land areas in a watershed. However, some watersheds will have loadings to the stream network from sources not associated with a land area. These are referred to as point sources. In order to account for the loadings from a point source, SWAT allows users to add daily or average daily loading data for point sources to the main channel network. These loadings are then routed through the channel network along with the loadings generated by the land areas. Procedures discussed for nitrogen can be applied to phosphorus as well. As mentioned earlier, HSPF model has similar provisions for point sources. SWAT computes chemical contributed by HRU in surface runoff, in lateral flow, chemicals leached below the soil profile, and chemical contributions in groundwater flow to stream.

B.2.10 SWMM (Storm Water Management Model)

B.2.10.1 Description

The U.S. EPA Storm Water Management Model is a comprehensive watershed model which is widely used for analysis of quantity and quality problems related to storm water runoff, combined sewers, sanitary sewers, and other drainage systems in urban areas, with many applications in non-urban areas as well, including floodplain hydraulics and analysis. SWMM simulates hydrographs and pollutographs (concentration vs. time) at any point in the drainage system. The model has been widely used for analysis of hydrologic and hydraulic problems of both combined and separate sewer systems as well as for urban non-point pollution problems. SWMM was originally developed by EPA in the period 1969-71 and was the first comprehensive model of its type for urban runoff analysis. The model performs best in urbanized areas with impervious drainage, which was its original intended application, but it has been widely used elsewhere. The greatest strengths of SWMM are in its ability to model the details of urban hydraulic systems, such as drains, detention basins, sewers, and related flow controls.

B.2.10.2 Model Type

SWMM is physically based distributed model. Single events and continuous simulations can be performed. Model inputs and outputs are presented in Table B-1.

B.2.10.3 Model Capabilities and Limitations

Spatial variation: The basic spatial unit for SWMM is the subcatchment, into which the modeled watershed is subdivided. SWMM is distributed model since the watershed can be divided into relatively homogeneous subcatchments. Data that characterize the subcatchment such as area, imperviousness, slope, roughness, width, depression storage, and infiltration parameters, land use can be input into the model.

Temporal variation: Single events and continuous simulations can be performed for any values of rainfall, runoff, and quality cycles for a watershed. However, the inter-storm interval is treated simplistically, the most significant processes being continued infiltration into a base flow, and build-up of contaminants on impervious surfaces. Its most common and successful application is to isolated storm events.

Model limitations: Technical limitations include lack of subsurface quality routing (a constant concentration is used), no interaction of quality processes (apart from adsorption), difficulty in simulation of wetlands quality processes (represented only as storage processes), and a weak scour deposition routine in the transport block. It is relatively difficult to use.

*Hydrology and water quality outputs: S*urface and subsurface runoff, flow routing through drainage network, storage and treatment can be simulated. SWMM can be used for planning and design. Planning mode is used for an overall assessment of urban runoff problem.

B.2.10.4 Procedures Used to Compute Model Essential Outputs

Runoff: Rainfall excess is calculated in SWMM by subtracting infiltration (based on Horton or Green and Ampt) and/or evaporation from precipitation. Rainfall excess is converted to runoff by coupling Manning's equation with the continuity equation. The newest version of SWMM also incorporates the runoff curve number method for estimating infiltration.

Sediment: Build-up and wash-off models are used in SWMM for pollutant loading. For impervious areas, a linear formulation is used to compute daily/hourly increases in particle accumulation. For pervious areas, a modified USLE determines sediment load.

Nitrogen: SWMM requires a percent land use for each subcatchment and an EMC associated with each land use. The combination of land use percentages and land use-based EMCs determines the concentration associated with each subcatchment. Event Mean Concentrations (EMCs) are applicable in both urban and agricultural environments. SWMM simulates hydrographs and (optionally) and pollutographs (concentration vs. time) at any point in the drainage system.

Phosphorus. Similarly EMC approach is used to compute to develop pullutographs (concentration vs. time)

Other pollutants: SWMM can simulate suspended solids (TSS), biological oxygen demand (BOD), chemical oxygen demand (COD),lead (Pb), cadmium (Cd), copper (Cu), and zinc (Zn). Event Mean Concentrations (EMCs) are calculated for each pollutant.

B.2.10.5 Model Availability and Simplicity for Users

SWMM is freely available to public. One of the limitations of SWMM is its difficulty for users. The biggest impediment to SWMM usage is the user interface, with its lack of menus and graphical output. Graphical interfaces are in the process of development.

B.2.10.6 History of Application

SWMM has an impressive longevity. It has been used in scores of U.S. cities as well as extensively in Canada, Europe, Australia and elsewhere. A large body of literature on theory and case studies is available, partly documented in a bibliography of SWMM-related publications and elsewhere. The model has been used for very complex hydraulic analysis for combined sewer overflow mitigation as well as for many storm water management planning studies and pollution abatement projects, and there are many instances of successful calibration and verification.

B.2.10.7 Runoff, Subsurface Flow, and Groundwater Flow

SWMM is designed to compute surface runoff from pervious and impervious area. SWMM also models snowmelt, subsurface drainage, and infiltration/inflow options. The subsurface drainage option is especially useful in locations where true overland flow rarely occurs because of flat, sandy soils.

B.2.10.8 Model Suitability for Onsite Wastewater Systems

SWMM simulates concentration vs. time of pollutants at any point in the drainage using event mean concentration approach. It doesn't simulate chemical transformation that occurs in the soil. This makes it a poor choice for decentralized wastewater systems; but it can be a good model for centralized wastewater systems where point source discharges are known.

B.2.11 WARMF (Watershed Analysis Risk Management Framework)

B.2.11.1 Description

WARMF is a decision support system sponsored by Electric Power Research Institute (EPRI) designed for watershed approach and TMDL calculation (Chen et al., 2001). WARMF calculates daily runoff, shallow groundwater flow, and water quality of a river basin. A river basin is divided into a network of land catchments, stream segments, and lake layers for hydrologic and water-quality simulations. WARMF is organized into five linked modules under one graphical user interface (GUI). The Engineering Module is a GIS-based watershed model that calculates daily runoff, shallow groundwater flow, hydrology, and water quality of a river basin. Land surface is characterized by land use and precipitation is deposited on the land catchments to calcuate snow and soil hydrology, and resulting surface runoff and groundwater accretion to river segments. Water is then routed from one river segment to the next, from river segments to reservoirs, and then from reservoirs to river segments, until the watershed terminus is reached. For chemical simulation, instead of using export coefficients, a complete mass balance is performed starting with atmospheric deposition and land application. Pollutants are routed with water in throughfall, infiltration, soil adsorption, exfiltration, and overland flow. WARMF provides several options for modeling reservoirs using 1D or 2D approaches. The Data Module contains meteorology, air quality, point source, reservoir release, and flow diversion data. It also contains observed flow and water quality data used for calibration. The data are accessed through the map-based interface and can be viewed and edited in both graphical and tabular format. The Knowledge Module stores supplemental watershed data, documents, case studies, or reports of past modeling activities for easy access by model users.

WARMF can simulate changes in stormwater runoff and water quality from urbanized areas and predict their impacts on the receiving waters that also receive runoff from non-urbanized areas within the drainage basin. WARMF includes algorithms used in SWMM for accumulation and wash off of pollutants in urban areas (Chen et al., 2001). Another important feature of WARMF is the integration of the 2D reservoir simulation model (CE-QUAL-W2) for in stream processes.

B.2.11.2 Model Type

WARMF is semi-empirical distributed and continuous simulation model. Model inputs and outputs are presented in Table B-1.

B.2.11.3 Model Capability, Strength, and Limitations

Spatial variation: Watersheds modeled by WARMF are divided into sub-basins. The percentage of land use type in each sub-basin can be defined to help discretization. However, unlike land use, the soil data are not imported into WARMF and soil parameters need to be adjusted for each sub-basin manually, which is a limitation. In SWAT and HSPF computations are made at HRU levels that are homogeneous with respect to land use and soil type. Generally, compared to other models such as SWAT, WARMF has a limited option for discretization.

Temporal variation: WARMF is a continuous model.

Model Limitations: GIS is internal, groundwater transport is weak, and soil database is generic, not site specific.

Hydrology and water quality outputs: WARMF can simulate stream flow, groundwater flow, lateral flow, sediment loading, nitrogen, phosphorus, heavy metals, and pesticides.

B.2.11.4 Procedures or Routines Used to Compute Model Essential Outputs

Runoff: A water balance approach is used to determine runoff. Precipitation and snowmelt will fall onto the ground surface. The water on impervious surfaces is subject to immediate runoff. The water on pervious surfaces may infiltrate, remain on the surface as detention storage, or flow as surface runoff. Surface water, which does not infiltrate into the soil may be ponded on the surface or run off as sheet flow. Detention storage is assumed to be a percentage of all the surface water on pervious surfaces.

Sediment: WARMF simulates the transport of clay, silt, and sand separately. The results are combined for total suspended sediment. The transport processes include the detachment of soil particles from the land surface, the suspension and deposition of detached soil particles in the overland flow, and the bed load transport on land. It is assumed that sediment transport does not occur when there is snow cover on the ground. It is also assumed that the groundwater flows do not carry sediments. The algorithms of sediment transport are based on ANSWERS. The first step in erosion is detachment. Detachment can be caused by kinetic energy of rainfall drops and the turbulence of overland flow. The sum of rainfall and turbulent detachment is compared to the bed load transport capacity. The detached soil in excess of transport capacity is immediately re-deposited. If the detached soil is less than transport capacity, all the detached soil will remain in suspension. Not all the water on the surface of a catchment flows out in a time step. The model mixes the suspended sediment in the total volume of the surface water. The runoff fraction will carry with it the average concentrations of the suspended sediment. The water retained on land surface will also have the same concentrations of suspended sediment for the next time step.

Nitrogen: A mass balance approach is used for calculating nutrient transport (Herr et al., 2001). The model extracts nutrients from soil according to net plant productivity specified in the input. The data are provided for each land use so that the model can calculate the terms for each land use and aggregate them for the catchment. A mass balance is conducted so the amount required for growth for each chemical component of the biomass is taken from the soil. WARMF simulates nitrogen uptake by plants, nitrification and denitrification. The plant uptake often removes more cations than anions from the soil solution. The model also simulates adsorption, and transport of nitrogen in soil layers.

Phosphorus: Phosphorus is adsorbed to soil particles. Phosphorus can exist in two forms, the dissolved fraction and the adsorbed fraction. The adsorption and desorption of phosphorus with the soil particles are reversible. The partition between dissolved and adsorbed fractions is a function of total suspended sediment. The adsorbed phosphorus settles with soil particles.

Pesticides and other organic contaminants: WARMF simulates different types of pesticide applications. The pesticide spray can land on the tree canopy or the soil. Pesticides that land on the tree canopy can be washed down to the soil as a part of the throughfall. Each pesticide can have different decay rates on the canopy, in the soil, and in the water.

B.2.11.5 Runoff, Subsurface Flow and Groundwater Flow

The model divides the catchments into layers. There can be up to five layers of soil in each catchment. Each soil layer has its own volumetric soil moisture content, horizontal and vertical hydraulic conductivity, field capacity, and saturated soil moisture content. A water balance is maintained for each layer. The model calculates infiltration into each layer, percolation from each layer to the layer below, lateral inflow into each layer from adjacent catchment, and lateral exfiltration from each soil layer to adjacent catchment, a stream, or a lake segment. If a catchment is adjacent to another catchment instead of a river or reservoir, the lateral flow out of the upstream segment will enter the downstream catchment (i.e., the lateral flow from an upstream catchment passes through the downstream catchments before joining the stream). However, if a catchment is adjacent to a river or a lake, the lateral flow enters the stream or lake. Infiltration into a layer depends on the potential vertical infiltration rate, which depends on moisture content of the layer and water available for infiltration to fill the void of the layer. The lesser of the two (the potential or the actual infiltration) is taken as the infiltration rate into the layer. The lateral exfiltration of water from soil layer is based on Darcy's Law which depends on horizontal hydraulic conductivity adjusted for moisture, temperature, and the hydraulic conductivity of any downstream catchment, slope of the catchment, width of the catchment parallel to its receiving stream, and thickness of the soil layer. The water balance is used to compute the soil moisture content of the layer based on infiltration to the layer, percolation out of the layer, lateral inflow from an upstream segment, evapotranspiration apportioned to layer; and exfiltration flow from layer. The model determines the new water volume, infiltration, percolation, and lateral flow. If the volume is greater than the volume of saturated soil moisture, the soil layer is already saturated. Darcy Law is used for groundwater transport between subcatchment or from subcatchment to the stream as base flow.

B.2.11.6 Model Suitability for Onsite Wastewater Systems

WARMF was adapted to simulate OWS loading directly to the subsurface. Algorithms were developed based on the findings of OWS research and scientific principles. The model accepts available data such as households served by OWS, number of people per household, characteristics of septic tank effluent, meteorology and land use. The model calculates the "edge-of-drain field pollution loads," rather than requiring them as input. In addition, the model predicts flow and water quality concentrations that can be verified by observed data. Septic system data from census database can be imported into WARMF. WARMF also has an option for grouping the type of septic system into standard, advanced and failing. The loading data for each type of land applied fertilizer or manure can be input on monthly basis (Herr et al., 2001). Other useful feature of WARMF is the TMDL module, which allows calculation of TMDLs for various management scenarios including septic systems and sewers.

As mentioned earlier, WARMF has similar provisions for point sources like SWAT and HSPF, discussed earlier. That is, if a sewer is used instead of decentralized OWS the discharge from wastewater treatment source can be considered as a point source discharge into the streams.

B.2.12 WMS (Watershed Modeling System)

B.2.12.1 Description

WMS is a graphical user interface (GUI), which integrates various models under the same umbrella and provides the linkage between them. WMS is a comprehensive graphical modeling environment for all phases of watershed hydrology and hydraulics. It supports hydrologic modeling with HEC-1 (HEC-HMS), TR-20, TR-55, Rational Method, NFF, MODRAT, and HSPF is now supported. Hydraulic models now supported in the WMS software include HEC-RAS and CE QUAL W2. 2D integrated hydrology (including channel hydraulics and groundwater interaction) can now be modeled with Girded Surface Subsurface Hydrologic Analysis (GSSHA). HEC1 is the most commonly used lumped parameter model designed to simulate surface runoff from a single precipitation event. TR-20 is designed to compute surface runoff from natural or synthetic rainstorm events. TR55 was developed by the NRCS as a simplified method to compute storm runoff in small, urbanized watersheds. MODRAT is the specialized Modified Rational Method program to compute surface runoff. The Storm Drain model does complete storm sewer network analysis for steady state or transient flow conditions. CE QUAL W2 is a 2D (profile) hydraulic model used for water quality analysis in rivers and reservoirs where vertical variation analysis is required. The Rational Method computes peak discharge from an area based on rainfall intensity and a runoff coefficient. The Hydrological Simulation Program - FORTRAN simulates hydrologic and water-quality processes on land surfaces, streams, and impoundments and was described in detail previously in this report. It is often used in the development of TMDLs. HEC-RAS is a 1D hydraulic model for computing water surface profiles for steady state or gradually varied flow. GSSHA is a distributed (2D) hydrologic model developed for analysis of surface runoff, channel hydraulics, and groundwater interaction. Water quality and sediment transport are also supported.

WMS is not a model so much as it is a convenient interface to standard models, including HEC-1, TR-20, NFF, and the Rational Method. HEC-HMS, TR-20, TR-55, Rational Method, NFF, MODRAT are not used for water quality modeling, hence they are not directly applicable to onsite wastewater systems and CE QUAL W2 is a 2D model that can be applied to in rivers and reservoirs and is included in other hydrologic models. The two models that can be used for OWS in WMS are HSPF and the newly developed GSSHA. HSPF has already been discussed. Some of the features of GSSHA is presented below.

GSSHA is a reformulation and enhancement of CASC-2D (Ogden and Julien, 2002). GSSHA can perform single event and continuous time simulations. It uses one and two dimensional diffusive wave flow routing at channels and overland planes respectively. Flow components of GSSHA is fully physically based whereas the sediment component is semi empirical. GSSHA considers other runoff generating mechanisms such as lateral saturated groundwater flow, exfiltration and stream/groundwater interaction. The sediment component is currently being reformulated based on physics based sediment transport concept. In its current version, the sediment transport formulation is based on USLE parameters. The nutrient modeling code was derived from the SWAT and the EPIC (Erosion/Productivity Impact Calculator) model (see SWAT nutrient transformation).

Table B-1. Watershed Models, Inputs, and Outputs.

Model	AGNPS	ANSWERS-2000	CREAMS-WT	GLEAMS
Inputs	SCS curve number, land slope, land slope shape factor, field slope length, channel slope, channel side slope, Manning's roughness coefficient, soil erodibility factor, cover and management factor, support practice factor, surface condition constant, aspect (direction of drainage), soil texture, fertilization level, fertilization availability factor, point source indicator, gully source level, chemical oxygen demand (COD) factor, impoundment factor and channel indicator	A raster-based DEM, soil coverage and a land use cover map Database files such as Soil IDs/ Particle Classes, Crop/Fertilizer Rotations, Drainage Coefficients, and Climate Station Locations	Rainfall, temperature and solar radiation, soil parameters such as saturated hydraulic conductivity, plant-available water, storage and porosity leaf are index are needed. Soil nitrogen and phosphorus concentrations, field capacity, and porosity. Concentrations of nitrogen and phosphorus in rainfall	Daily rainfall and temperature; monthly temperature, solar radiation, wind movement, and dew point temperature; soil data; pesticide characteristics and application data; fertilization and tillage data
Outputs	Hydrology outputs (runoff volume, peak runoff rate, fraction of runoff generated within the cell), Sediment (by particle size and in total), sediment yield, sediment concentration, sediment particle size distribution, upland erosion, channel erosion, amount of deposition (%) sediment generated within the cell (tons), delivery ratio. Nutrient outputs include nitrogen, and phosphorus sediment associated mass (lbs/ac), concentration (ppm), mass of soluble material in runoff (lbs/ac), Chemical Oxygen Demand (ppm)	Daily or cumulative values, by cell or channel outlet, Runoff (mm), Sediment (kg/ha), NO3 (kg), Dissolved NH4 (kg), Sediment adsorbed NH4 (kg), Dissolved PO4 (kg), Sediment adsorbed PO4 (kg), Sediment adsorbed TKN (kg).	Runoff, percolation, erosion, and dissolved and adsorbed plant nutrients and pesticides	Daily, monthly, annual. runoff, sediment, pesticide mass and concentration, percolation volume, plant nutrient mass and concentrations.

Table B-1 Continued.

Model	HSPF	MANAGE	MIKE SHE	MODFLOW
Inputs	Time series inputs such as precipitation, air temperature, dew point temperature, wind movement, solar radiation, evaporation/evapotranspiration and upstream inflows, lateral sediment inputs, upstream or tributary inflows, and external constituent mass loadings. Air temperature, wind, solar radiation, humidity, cloud cover, tillage practices, point sources, and (or) pesticide applications Physical measurements such as land area, channels, and reservoirs	GIS coverages such as land use, soils, sewers, public water systems, community water supply wells, roads are used to generate data for MANAGE model. Data extracted from GIS coverages used for computing model outputs	Digital Elevation Models (DEMs), Daily precipitation and temperature, Land Cover Database (NLCD), vegetation growth and consumptive use characteristics and average annual watering rates, GIS stream coverages, dam operational and physical information, wastewater discharge data, soil data, hydrologic soil groups, geologic data, hydraulic conductivity, groundwater Surface Water Interaction (Leakage coefficients and the magnitudes of gaining and losing reaches) , groundwater withdrawals, and boundary conditions.	Initial conditions, hydraulic properties, and stresses must be specified
Outputs	Time histories of sediment loads, runoff rates, and nutrient and chemical concentrations Quantity and quality of runoff from an urban or agricultural watershed, flow rate, sediment load, and nutrient and pesticide concentrations are predicted.	Hotspot mapping -a rapid screening of potential high-risk areas; watershed indicators - measures of generalized ecosystem health based on soil and land use characteristics. Locates high-infiltration areas and runoff-generating soils using soil permeability and water table depth. High intensity land use where pollutants are most likely to be generated Hotspots where pollutants are typically generated Hotspots for storm water runoff , septic system failure, risks in groundwater	Simulates surface runoff, sediment loading, nitrogen, phosphorus and pesticides. Overland flow, unsaturated flow, vegetation-based evapotranspiration, and groundwater flow are predicted.	Head is Primary output. Other output includes the complete listing of all input data, draw down, and budget data. Budget data are printed as a summary in the listing file, and detailed budget data for all model cells can be written into a separate file.

Table B-1 Continued.

Model	SWAT	SWMM	WARMF	WMS
Inputs	GIS data layers of elevation (DEM), soils, and land use are used to generate the input files. Observed or model generated temperature and precipitation Subbasin parameters such as slope, slope length, the stream network generated from DEM. Characteristics of the stream network (channel slope, length, and width) are all derived from the DEM. The SWAT model has an internal database of soil properties COOP weather data are available from the NOAA (National Oceanic and Atmospheric Administration). Agricultural management scenarios such as fertilizer application rates, tillage types, harvesting operations can be input in to the model.	Area, imperviousness, slope, roughness, width (a shape factor), depression storage, and infiltration parameters for either the Horton or Green-Ampt equations Precipitation, and evaporation, snowmelt, subsurface drainage, and infiltration/inflow, pollutant concentration. Most output is tabular	GIS coverages such as land use, soils, sewers, public water systems, community water supply wells, political boundaries and roads.	WMS is user interface for other models Model input depends on model component used.
Outputs	SWAT predicts various hydrology and water quality outputs on daily basis. Some of SWAT predicted out puts include, surface runoff, groundwater contribution to flow, sediment yield, nutrient yield and biomass	Hydrographs and pollutographs (concentration vs. time) Flow depths and velocities Additional quality output includes loads, source identification, continuity, and residuals (e.g., sludge)	WARMF can simulate surface runoff, sediment loading, nitrogen, phosphorus and heavy metals, pesticides.	Depends on model component used

Table B-2. Watershed Model Types, Capabilities, and Limitations.

Model	AGNPS	ANNAGNPS	ANSWERS-2000	CREAMS-WT	GLEAMS
Model type	Field/ Watershed Agricultural Watershed	Same as AGNPS	Small Watershed Agricultural Watershed	Field Scale	Field scale
Purpose	Simulates runoff, sediment, and nutrient transport primarily from agricultural watersheds	Same as AGNPS	Simulates runoff, erosion, nutrients and effectiveness of BMPs in reducing sediment and nutrients.	Simulates nutrients, pesticides, and sediment.	Simulates nutrients, pesticides, and sediment
Domain	Public	Public	Public	Public	Public
Level of analysis	Screening Detailed	Screening Detailed	Screening Detailed	Screening Detailed	Screening Detailed
Model processes	Empirical	Empirical	Physically based	Semi-Empirical	Semi-Empirical
Spatial and Temporal	Distributed Event	Distributed Continuous	Distributed Continuous	Lumped Continuous	Lumped Continuous
Level of skill	Medium to high	Medium to high	Medium to high	Medium	Medium
Model limitations	No day to day tracking of sediment attached chemicals deposited in stream Considers only surface water and infiltration but not subsurface flow Areal extent limited by the assumption of spatially uniform distributed rainfall Not specifically set up for DWTS input	Not specifically set up for DWTS input	Not good for large watershed and long simulations Nutrient transformations and transport relies on the empirical equations. Doesn't work equally good for all land uses and soil types	Applicable to field size, homogeneous areas, Stream flow is only from runoff	Good to field size, homogeneous areas, Better subsurface and leaching routines but stream flow is only from runoff

Table B-2 Continued.

Model	HSPF	MANAGE	MIKE SHE	MODFLOW
Model type	Watershed Agricultural and Urban	Watershed	Watershed-Groundwater Agricultural	Field Watershed
Purpose	Simulates nutrients, pesticides, and sediment	GIS – Pollutant Mass Loading model to assess relative effects	Integrated groundwater, surface water vadose zone, and geochemistry.	Groundwater Detailed groundwater, recharge, and pollutant transport in combination with other models
Domain	Public		Proprietary	Public
Level of analysis	Screening Detailed	Screening	Detailed	Detailed
Model processes	Physically based		Physical	Physical
Spatial and Temporal	Distributed Continuous		Distributed Event and continuous	Distributed
Level of skill	Medium to high	Low	High	Medium
Model limitations	Good for in-stream processes, weak for upland (overland) processes, No specific module for DWTS input Groundwater transport is weak.	Screening model only, does not provide account for many hydrologic and transport processes	High complexity and difficult to use Not in public domain. Relatively expensive to implement. Input and calibration data requirements are very large.	Representation of overland process and river flow is poor.

Table B-2 Continued.

Model	SWAT	WARMF	WMS
Model type	Watersheds Agricultural	Watershed	Watershed Agricultural Urban
Purpose	Predicts water, sediment and chemical yields	Specific sediment and chemical yields	Predicts, flow, sediment, nutrients
Domain Platform	Public	Public	Proprietary
Level of analysis	Screening Detailed	Screening Detailed	Screening detailed
Model processes	Semi Empirical	Semi Empirical	Has both types
Spatial and Temporal	Distributed Continuous	Distributed Continuous	Has both distributed and lumped models
Level of skill	Medium	Medium	Medium to High
Model limitations	HRUs may not be spatially contingent No interaction between HRUs Uses unvalidated assumptions for in- stream-processes Stream process algorithms are poor Not specifically set up for DWTS input, must use other methods, groundwater transport is weak	GIS is internal, not typical ArcGIS, current cost of implementation is high, groundwater transport is weak, Soil data base is generic , not site specific	

B.3 Conclusions

Most of the models reviewed to date could be used for some application involving OWS. One notable conclusion of this review is that most models are designed primarily for simulating either surface-water hydrology and pollutant transport or groundwater hydrology and pollutant transport, but not both. Surface-water models usually account for movement and reactions of water and chemicals in the soil but consider groundwater flow to streams in a simple manner. Typically, these models do not allow for output of groundwater concentrations, although the chemical mass in groundwater is tracked, and thus concentrations could be determined indirectly. Some surface water models were designed to simulate only storm events. These models are clearly not appropriate for OWS applications.

The groundwater models reviewed did not rigorously represent surface-water hydrology, or pollutant transport in surface water. Rather, streams are usually treated as a sink or source of water and chemicals to the groundwater system and the combined effects of precipitation, evapotranspiration, snowmelt, and runoff are simulated as a single recharge term to the aquifer in the model. However, when aquifer protection is the goal, these models would be necessary in order to accurately simulate relevant parameters and processes such as pumping, heterogeneities, temporal changes in hydraulic gradients, and mixing of infiltrating water (with the chemical load) and aquifer flow.

The only model reviewed that is capable of simulating most of the relevant processes in a completely integrated surface and groundwater is MIKE-SHE, which can also rigorously simulate vadose-zone transport and pollutant transport in all compartments. However, this model is complex in that the input-data requirements are very large, and a sophisticated user would be required. The model would undoubtedly be very difficult to calibrate and the large number of linked equations would cause the model performance to be "stiff," or require numerical-performance input parameters to be within a very small range that is likely to vary for different simulations. It also is not commonly used or publicly available in the United States. The versatility of this model is likely to be useful for problems where surface-water / groundwater interactions at the watershed scale are thought to be critically important, or where tracking concentrations in both stream and groundwater spatially is required. However, it would not be an efficient choice for the majority of problems where either groundwater or surface water is of primary concern, and is not likely to fit within the budget for most watershed management projects.

The most efficient and cost-effective approach in most cases will be to use a groundwater model for groundwater issues, and a surface-water model with limited groundwater capabilities for problems where surface-water quality predictions are important (e.g., TMDLs). In the realm of groundwater models, ANNGNPS, CREAMS-WT and GLEAMS have routines to simulate subsurface flow and leaching but subsurface flow and leaching do not contribute to stream flows. Although these models can be good to simulate the effect of OWS pollutants on groundwater, they are not suited to simulate flow, transport, and reactions in aquifers. Thus, for groundwater models, the MODFLOW suite is recommended because of the wide use, public availability, confidence in the model by the modeling and regulatory community, and excellent documentation. Depending on the contaminants of interest, MT3D or RT3D are modules associated with MODFLOW that can handle chemical transport and reactions. However, a weakness of these models is that chemical mass loading or concentrations must be input directly to the water table. Thus, it may be useful to combine the leaching capabilities of models such as ANNGNPS, CREAMS-WT, and GLEAMS with MODFLOW to allow more rigorous input of chemical concentrations to the top of the aquifer. In fact, SWAT has been coupled with MODFLOW (Perkins and Sophocleous, 1999; Sophocleous et al., 1999() for this purpose. However, this code is not publicly available at this time, and the link was performed only for physical hydrologic problems, not pollutant transport problems.

For surface-water quality problems, more options exist. It is critically important that the model be able to allow pollutant movement and transformation through the subsurface to streams because this is the primary means for OWS contributions to streams. Several of the models reviewed would be appropriate for use, including, HSPF, SWAT, WARMF, and GSSHA. HSPF, SWAT, and WARMF can all be implemented through the EPA BASINS system, which is essentially a graphical user interface (GUI) for a variety of watershed models. BASINS can be accesses via the internet at no cost and has many users. Because the EPA BASINS system is publicly available and readily accepted by the technical and regulatory community, the models associated with this system are preferred by the authors of this report.

WARMF is unique among these three because it contains modules specifically for OWS inputs (flows and concentrations, number of systems per sub-watershed, various treatment options, etc), TMDLs, and geochemistry. SWAT and HSPF have been used to simulate TMDL problems, but the TMDL module with its scenario management options and graphics facilitates the process. WARMF allows input of user-specified geochemical equations that may be very useful for phosphorus reactions or virus retention. Perhaps the most serious advantage of WARMF is that OWS effluent can be injected as a liquid into the shallow subsurface, with known flow rates and concentrations. SWAT and HSPF would require simulating OWS input as fertilizer application, which is problematic either precipitation or irrigation water must be relied upon for carrying the "OWS effluent" downward. Both options are problematic. Precipitation would cause a sporadic input, unlike continuously flowing OWS, and irrigation water may run off the land surface, incorrectly adding to stream flow. Both present problems for mass-balance tracking of OWS pollutants. Another advantage of WARMF is that the GIS system used for data input is self contained, thus users do not need to purchase ARC-GIS system to populate the model., but can still use the existing GIS-based governmental watershed databases This is primarily an advantage for regulators and planners who do not use GIS. This same feature can also be viewed as a disadvantage for professional hydrologists and engineers, because they are more familiar with ARC GIS systems, and likely already own the software. Another important disadvantage of WARMF is that soils information, perhaps the most onerous of the input-date requirements, cannot be directly input from GIS-based soil databases. The soil properties must be manually input by the user in one or more soil layers. However, an alternate positive view on this issues is that it allows for a more scrutinous evaluation of soil data input, especially below the surface (the typical soil databases are generally applicable to the top few feet of the land surface).

WARMF is new to the BASINS system, however, and SWAT and HSPF have many more users. Thus, our goal is to implement WARMF and one of SWAT and HSPF for each watershed in this study. Of the latter two models, SWAT, in our opinion, is better at simulating land-based processes. Because OWS and most competing inputs are land based, SWAT is a better choice for OWS applications.

APPENDIX C

Important Documents Associated with this Guide

Adamus, C.L., Bergman, M.J. 1995, Estimating nonpoint-source pollution loads with a GIS screening model: Water Resources Bulletin, v. 31, no. 4, p. 647-655. http://www.awra.org/jawra/

Bossong, C., Caine, J., Stannard, D., Flynn, J., Stevens, M., Heiny-Dash, J. 2003. Hydrologic Conditions and Assessment of Water Resources in the Turkey Creek Watershed, Jefferson County, Colorado, 1998-2001. USGS Water-Resources Investigations Report 03-4034. 140 p. http://pubs.usgs.gov/wri/wri03-4034/

Chen, C.W., Herr, J., Weintraub, L.H.Z. 2001. Watershed analysis risk management framework: Update One: A decision support system for watershed analysis and total maximum daily load calculation, allocation and implementation, EPRI, Palo Alto, CA. http://mydocs.epri.com/docs/public/000000000001005181.pdf

Geza, M., McCray, J.E. 2008. Effects of soil data resolution on SWAT model stream flow and water quality predictions. Journal of Environmental Management, Volume 88, Issue 3, August 2008, Pages 393-406. http://www.sciencedirect.com/science/journal/03014797

Geza M., Poeter E.P., McCray J.E. 2008. Hydrologic-Model Parameter Selection and Estimation for a Small Mountain Watershed. J. of Hydrology, (in press). www.elsevier.com/locate/jhydrol OR mgezanis@mines.edu

Geza, M., McCray, J.E. 2008. User guide for modeling watershed-scale impacts of onsite wastewater systems: Case studies of impacts of onsite systems in Turkey Creek Watershed, Colorado. NOWRA Annual Conference, Memphis, TN, April 7 - 10, 2008. http://www.nowra.org/files/Nitrogen%20Symposium%20presentations/Geza.pdf

Geza M., McCray J.E., Murray K.E. 2008. Phosphorus Parameter Sensitivity, Calibration and Evaluation of Impacts Onsite Wastewater Systems in Turkey Creek Watershed, Colorado (in review) mgezanis@mines.edu OR klowe@mines.edu

Geza M., McCray J.E., Murray K.E. 2008. Evaluation of Nitrate Parameter Sensitivity, Calibration and Impacts of Onsite Wastewater Systems in Turkey Creek Watershed, Colorado (in review) mgezanis@mines.edu or klowe@mines.edu

Heatwole, K.K., McCray, J.E. 2007. Modeling potential vadose-zone transport of nitrogen from onsite wastewater systems at the development scale, J. Contaminant Hydrology, 91, (1-2), 184-201. www.elsevier.com/locate/jconhyd

Herr J.L. Weintraub, C.W. Chen. 2001. User's Guide to WARMF. Documentation of Graphical User Interface. Prepared for Electric Power Research Institute, Palo Alto, CA. http://www.epa.gov/athens/wwqtsc/html/warmf.html

Hill, M.C., Tiedeman, C.R. 2007, Effective groundwater model calibration, with analysis of sensitivities, predictions, and uncertainty: New York, Wiley and Sons, 455 p.

Hinkle, S.R., Böhlke, J.K., Duff, J.H., Morgan, D.S., Weick, R.J. 2007, Aquifer scale controls on the distribution of nitrate and ammonium in groundwater near La Pine, Oregon, USA: Journal of Hydrology, v. 333, p. 486-503. www.elsevier.com/locate/jhydrol

Hinkle, S.R., Morgan, D.S., Orzol, L.L., Polette, D.J. 2007, Groundwater redox zonation near La Pine, Oregon – Relation to river position within the aquifer-riparian zone continuum: U.S. Geological Survey Scientific Investigations Report 2007-5239, 28 p.; last accessed November 9, 2007, at http://pubs.usgs.gov/sir/2007/5239/.

Hinkle, S.T., Weick, R.J., Johnson, J.M., Cahill, J.D., Smith, S.G., Rich, B.J. (2005). Organic wastewater compounds, pharmaceuticals, and coliphage in groundwater receiving discharge from onsite wastewater treatment systems near La Pine, Oregon: Occurrence and implications for transport. *U.S. Geological Survey Scientific Investigations Report 2005* 5055: 98. www.elsevier.com/locate/jhydrol

Kellogg, D.Q., Gold, A.J. 1996, Technical documentation, nutrient loading component of the MANAGE Geographic Information System-based risk assessment method: University of Rhode Island Cooperative Extension. http://www.uri.edu/ce/wq/NEMO/Tools/pollution_assessment.htm#manage

Lemonds, P.J., McCray, J.E, 2003. Modeling phosphorus transport in the Blue River Watershed, Summit County, Colorado, and Proceedings of the First Interagency Conference on Research in the Watersheds, U.S.D.A, U.S Geol. Survey, Bureau of Land Manag., and Natural Resour. Conservation Serv., U.S. E.P.A, U.S. Forest Serv., Consortium of Universities for the Advancement of Hydrologic Science, October 28-30, 2003, Benson, AZ. http://www.tucson.ars.ag.gov/icrw/Proceedings/Lemonds.pdf

Lin Z., Radcliffe D.E. 2006. Automatic Calibration and Predictive Uncertainty Analysis of a Semi distributed Watershed Model. Vadose Zone Journal 5:248-260. http://vzj.scijournals.org/

McCray J.E., Kirkland, S.L., Siegrist, R.L., Thyne, G.D.. 2005. Model Parameters for Simulating Fate and Transport of Onsite Wastewater Nutrients.Vol. 43, No. 4. GROUND WATER, July-August 2005 (pages 628-639). http://www.blackwellpublishing.com/journal.asp?ref=0017-467X&site=1

McCray J.E., Geza M., Poeter, E.P. Hydrologic and Chemical-Transport Modeling to Assess Watershed-Scale Impacts of Onsite Wastewater Pollutants and Future Scenarios. ASABE, 21st Century Watershed Technology: Improving Water Quality and Environment March 29-April 3, 2008 University of Concepcion, Concepcion, Chile.

http://asae.frymulti.com/conference.asp?confid=iwqe2008

Morgan, D.S., Hinkle, S.R., Weick, R.J. 2007, Evaluation of approaches for managing nitrate loading from onsite wastewater systems near La Pine, Oregon: U.S. Geological Survey Scientific Investigations Report 2007-5237, 64 p.; also available online at http://pubs.usgs.gov/sir/2007/5237/

NCDC, 2007. The National Oceanic and Atmospheric Administration (NOAA). National Climatic Data Center. US Department of Commerce. http://www.ncdc.noaa.gov/oa/climate/stationlocator.html

Neitsch, S.L., Arnold, J.G., Kiniry, J.R., Williams, J.R., . 2002. Soil and Water Assessment Tool (SWAT). Theoretical documentation, version 2005. http://www.brc.tamus.edu/swat/doc.html

Poeter, E.P., Anderson, D. 2005, Multi-model Ranking and Inference in Ground-Water Modeling, Ground Water v. 43, no. 4: 597-605. http://www.blackwellpublishing.com/journal.asp?ref=0017-467X&site=1

Poeter, E.P, Hill M.C., Banta E.R., Mehl S., Christensen S. 2005. UCODE_2005 and Six Other Computer Codes for Universal Sensitivity Analysis, Calibration, and Uncertainty Evaluation: U.S. Geological Survey Techniques and Methods 6-A11(2005) 283 pp. http://water.usgs.gov/software/ or http://www.mines.edu/igwmc/freeware/ucode/UCODE_2005_update_2008-02-10.pdf

Refsgaard, J.C. 1997. Parameterization, calibration and validation of distributed hydrologic models. Journal of Hydrology 198, 69-97. www.elsevier.com/locate/jhydrol

Tucholke, M.B., McCray, J.E., Thyne, G.D., Waskom, R.M. (2007). Correlating Denitrification Rates to Soil Texture using Hierarchical Cluster Analysis. Eleventh National Symposium on Individual and Small Community Sewage Systems, ASABE, October 2007. mtucholk@mines.edu, OR jmccray@mines.edu

USDA, 2007. Natural Resources Conservation Service (NRCS), National Cartography and Geospatial Center. Soil Survey Geographic (SSURGO). http://www.ncgc.nrcs.usda.gov/products/datasets/statsgo/index.html

Valiela, I., Collins, G., Kremer, J., Lajtha, K., Geist, M., Seely, B., Brawley, J., Sham, C.H. 1997, Nitrogen loading from coastal watersheds to receiving estuaries: New method and application: Ecological Applications, v. 7, no. 2, p. 358-380. http://ciceet.unh.edu/news/releases/reportRelease/pdf/valiela_final_report.pdf

Weintraub L.H.Z., Chen, C.W., Goldstein, R.A., Siegrist, R.L. 2004. WARMF: A Watershed Modeling Tool For Onsite Wastewater Systems. In Proceedings of The Tenth National Symposium on Individual and Small Community Sewage Systems, 636-646. ASAE. St. Joseph, MI. Presented March 21-24, 2004, Sacramento, CA. http://www.systechengineering.com/downloaddocs/OWS_WEF_WS2004.pdf

References

Adamus, C.L., Bergman, M.J. 1995. Estimating non-point source pollution loads with a GIS screening model: Water Resources Bulletin, v. 31, no. 4, p. 647-655.

Ajami, N.K., Gupta, H., Wagner, T. Sorooshian, S. 2004. "Calibration of a Semi Distributed Hydrologic Model For Streamflow Estimation Along A River System," Journal of hydrology, 298,112-135.

Anderson, M.P., Woessner, W.W. 1992, Applied groundwater modeling: Simulation of flow and advective transport: San Diego, CA, Academic Press, xviii, 381 p.

Arnold, J.G., Srinivasan, R., Muttiah, R.S., Williams, J.R. 1998. Large area hydrologic modeling and assessment. part I: Model development. J. Amer. Water Resour. Assoc. 34(1): 73-89.

Arnold, J.G. 1992: Spatial Scale Variability in Model Development and Parameterization: Ph.D. Dissertation, Purdue University, West Lafayette, IN, 183 p.

Beasley, D.B., Huggins, L.F., Monke, E.J. 1980, ANSWERS: A model for watershed planning: Transactions of the ASAE, v. 23, no. 4, p. 938-944.

Bedessem, M.J., Edgar, T.V. Roll, R. 2005. Nitrogen removal in laboratory model leachfields with organic-rich layers. J. Environmental Quality, 34(3): 936-942.

Beven, K.J., Freer, J. 2001. Equifinality, data assimilation, and uncertainty estimation in mechanistic modeling of complex environmental systems using the GLUE methodology. Journal of Hydrology 249, 11-29.

Beven, K.J., Binley, A.M. 1992. The future of distributed models: model calibration and uncertainty prediction. Hydrological Processing 6, 279-298.

Bissett, L., Poeter, E.P. 1994: Field assessment of stream/aquifer interaction under semi-arid conditions and problems with computer representation. Completion Report No: 185, Colorado Water Resources Research Inst., Colorado State University.

Borah, D.K., Yagow, G., Saleh, A., Barnes, P.L., Rosenthal, W., Krug, E.C., Hauck, L.M. 2006, Sediment and nutrient modeling for TMDL development and implementation: Transactions of the ASABE, v. 49, no. 4, p. 967-986.

Borrero, C., Pena, F. Torrent, J. 1988. Phosphate sorption by calcium carbonate in some soils of the Mediterranean part of Spain. Geoderma, 42: 261-269.

Bossong, C., Caine, J., Stannard, D., Flynn, J., Stevens, M., Heiny-Dash, J. 2003. Hydrologic Conditions and Assessment of Water Resources in the Turkey Creek Watershed, Jefferson County, Colorado, 1998-2001. USGS Water-Resources Investigations Report 03-4034. 140 p.

Bouraoui, F., Dillaha, T.A. 1996, ANSWERS-2000: Runoff and sediment transport model: Journal of Environmental Engineering, v. 122, no. 6, p. 493-502.

Boyle, D.P., Gupta, H.V., Sorooshian, S., Koren, V., Zhang, Z., Smith, M. 2001. Toward improved stream flow forecast: value of semi distributed modeling. Water Resources Research 37(11), 2749-2759.

Bradley, C Brown, A.G, 1997: Modeling of hydrological processes in a floodplain wetland.Groundwater/ surface water ecotones: biological and hydrological interactions and management options (Gibert, J; Mathieu, J; Fournier, F, eds), pp. 102-110.

Brady, N.C., Weil, R.R. 2002. The nature and properties of soils. Pearson Education Inc., Upper Saddle River, New Jersey, 960 pp.

Brandes, M., Chowhdry, N.A., Cheng, W.W. 1974. Experimental study on removal of pollutants from domestic sewage by underdrained soil filters, Proceedings of the national home sewage disposal dymposium, Chicago, IL, pp. 29-36.

Bredehoeft, J.D., Konikow, L.F., Ground-water models: Validate or invalidate, Ground Water, 31(2), 178-179.

Buetow, W. S. 2002. Onsite Wastewater Nitrogen Contributions to a Shallow Aquifer and Adjacent Stream. MS thesis. Soil Science Dept., College of Agriculture and Life Science, North Carolina State University, NC.

Caraco, N.F., Cole, J.J. 1999. Human impact on nitrate export: an analysis using major world rivers. Ambio 28, 167-70.

Chapuis-Lardy, L., Fiorini, J., Toth, J., Dou, Z. 2004. Phosphorus concentration and solubility and dairy feces: variability and affecting factors. Journal of Dairy Science, 87: 4334-4341.

Chamberlin, T.C. 1890. The method of multiple working hypotheses. Science 15: 93?98. Reprinted 1965, Science, 148: 754-759.

Chen C.W., Herr, J., Laura H., Z. Weintraub, Z. 2001. Watershed Analysis Risk Management Framework: Update One: A Decision Support System for Watershed Analysis and Total Maximum Daily Load Calculation, Allocation and Implementation, EPRI, Palo Alto, CA: 2001. 1005181.

Christiaens, K., Feyen, J.: Use of sensitivity and uncertainty measures in distributed hydrological modeling with an application to the MIKE SHE model, Water Resour. Res., 38, 1169, doi:10.1029/2001WR000 478, 2002.

Clement, T.P., Sun, Y., Hooker, B.S., Petersen, J.N. 1998. Modeling multi-species reactive transport in groundwater. Groundwater Monitoring and Remediation 18 (2), 79-92.

Clement, T.P. 1997. RT3D – A Modular Computer Code for Simulating Reactive Multi-Species Transport in 3-Dimensional Groundwater Aquifers. Pacific Northwest National Laboratory, Richland, WA, USA. PNNL-11720. Found online at: http://bioprocess.pnl.gov/rt3d.htm.

Codd, G.A. 1995. Cyanobacterial toxins: occurrence, properties and biological significance. Water Science and Technology, 32(4): 149-156.

Cohen, S. 1996: Pesticides in groundwater in the United States: Monitoring, modeling, and risks from the U.S. perspective. J. Environ. Sci. Health, Part B: Pest., Food., Contamn., Ag.Wastes, B31, pp. 345-352.

Cole, J.J., Peierls, B.L., Caraco, N.F., Pace, M.L. 1993. Nitrogen loading of rivers as a human-driven process. In: McDonnell, M.J., Pickett, S.T.A. (Eds.), Humans as Components of Ecosystems: The Ecology of Subtle Human Effects and Populated Areas. Springer-Verlag, New York.

Cooley, R.L., Naff, R.L. 1990. Regression modeling of groundwater flow: U.S. Geological Survey Techniques of Water-Resources Investigations. Book 3, Chapter B4, 232 p.

Corbett, C.W.; M. Wahl, D. Porter; D. Edwards, C. Moise, 1997: Non-Point Source runoff modeling. A comparison of forested watershed and urban watershed on the South

Corwin D.L. 1996. Applications of GIS to the modeling of non-point pollutants in the vadose zone. Corwin D.L (ed), ASA-CSSA-SSA, Madison, Wisconsin, USA. Special publication N° 48, pp. 69-100.

Costa, J.E., Howes, B.L., Janik, D., Aubrey, D., Gunn, E., Giblin, A.E. 1999. Managing Anthropogenic Nitrogen Inputs to Coastal Embayments: Technical Basis of a Management Strategy Adopted for Buzzards Bay. Buzzards Bay Proj. Tech. Rep

Craig, J.P.; Weiss, R.C. 1993: Use of the GLEAMS model to estimate pesticide overland and subsurface transport in USDA forest service nursery applications. Proceedings of the IAWQ 1st International Conference on Diffuse (Non-point) Pollution: Sources, Prevention, Impact, Abatement, 28 (3-5), pp 425-42

Dabral S., Cohen M. 2001.ANSWERS-2000 Areal Non-point Source Watershed Environment Response Simulation with Questions Graphical User Interface ABE6254 – Simulation of Small Agricultural Watersheds Project Report 1. Virginia Tech.

Danish Hydraulic Institute (DHI). 1998. MIKE SHE Water Movement – User Guide and Technical Reference Manual, Edition 1.1.

Dettmann, E. 2001. Effects of water residence time on annual export and denitrification of nitrogen in estuaries: A model analysis. Estuaries 24, 481-490.

Diab, S., Shilo, M. 1988. Nitrogen transformation in wastewater reclamation. Water Res., 22(5): 557-563.

Dillaha, T.A., Wolfe, M.L., Shirmohammadi, A., Byne, F.W. 2001. ANSWERS-2000. Non-point source water quality models: Their use and application. Final Report of USDA-CSREES southern region research project S-273 Development and Application of Comprehensive Agricultural Ecosystems Models.

Doherty, J. 2004. PEST – Model-independent parameter estimation user's manual. 5th ed. Watermark Numerical Computing, Brisbane, Australia.

Donigan, A.S., Imhoff, J.C., Bicknell, B.R., Kittle, J.L. 1984, Application guide for hydrological simulation program: FORTRAN (HSPF), U.S. EPA-600/3-84-065, p. 177.

Donigian, A.S. Jr., Bicknell, B.R., Imhoff, J.C., 1995: Hydrological Simulation Program – FORTRAN. Chap. 12, Computer Models of Watershed Hydrology (V. Singh, ed.), pp.395-442. Highlands Ranch, CO: Water Resources Publications.

Donigian, A.S., Jr., Huber, W.C. 1991. Modeling of Non-Point Source Water Quality in Urban and Non-Urban Areas. EPA/600/3-91/039. U.S. Environmental Protection Agency, Environmental Research Laboratory, Athens, GA. 72 p.

Drewes, J.E., Heberer, T., Rauch, T., Reddersen, K. (2003). Fate of pharmaceuticals during groundwater recharge. *Ground Water Monitoring and Remediation* 23(3): 64-72.

Duan, Q., Sorooshian, S., Gupta, V.K. 1992. Effective and efficient global optimization for conceptual rainfall–runoff models. Water Resources and Research 28 (4), 1015-1031.

Eckhardt, K., Arnold, J.G. 2001. Automatic calibration of distributed catchment model. Journal of Hydrology 251, 103-109.

Engel B., Storm, D., White, M., Arnold, J.G. 2007. A hydrologic/water quality model application protocol. J. American Water Resources Assoc. (in press).

EPA, 2002. Onsite wastewater treatment systems manual. In: O. Cincinnati, Office of Water, Office of Research Development (Editor).

Evans, B.M., Sheeder, S.A., Corradini, K.J. 2006, AVGWLF Version 6.3 Users Guide: Penn State Institutes of the Environment, The Pennsylvania State Universityp. 79.

Fishman, G.S. 1996. Monte Carlo: Concepts, Algorithms and Applications. Springer, New York, U.S.

Foerster, J., Milne-Home, W.A. 1995. Application of AGNPS to model nutrient generation rates under different farming management practices at the Gunnedah Research Centre catchment. AUST. J. EXP. AGRIC. ; vol. 35, no. 7, pp. 961-967.

Fredericks, J., Labadie, J. 1993: Decision support system for conjunctive stream-aquifer management in the South Platte Basin. Seeking an integrated approach to watershed management in the South Platte Basin (K. Klein , D. Williams, eds). Colorado Water Res. Res. Inst. Inform. Ser. 76, pp. 57-58. Colorado State University.

Foster, G.R., Lane, L.J., Nowlin, J.D., Laflen, J.M., Young, R.A. 1980. A Model to Estimate Sediment Yield from Field-Sized Areas: Development of the Model. In: Knisel, W.G. (ed.) 1980. CREAMS: A Field Scale Model for Chemicals, Runoff, and Erosion from Agricultural Management Systems, USDA, Cons. Research Report No. 26, pp. 36-63.

Gerritse, R.G. 1993. Prediction of travel times of phosphate in soils at a disposal site for wastewater. Water Research, 27(2): 263-267.

Geza M., Poeter E.P., McCray J.E. 2008. Hydrologic-Model Parameter Selection and Estimation for a Small Mountain Watershed. J. of Hydrology, (in press).

Geza, M., McCray, J.E. 2008. User guide for modeling watershed-scale impacts of onsite wastewater systems: Case studies of impacts of onsite systems in Turkey Creek Watershed, Colorado. NOWRA Annual Conference, Memphis, TN, April 7 - 10, 2008. http://www.nowra.org/files/Nitrogen%20Symposium%20presentations/Geza.pdf

Geza, M., McCray, J.E. 2008. Effects of soil data resolution on SWAT model stream flow and water quality predictions. Journal of Environmental Management, Volume 88, Issue 3, August 2008, Pages 393-406.

Goss, D. W. 1992: Screening procedure for soils and pesticides for potential water quality impacts. Weed Technology, 6 (3), pp. 701-708.

Grunwald, S., Norton, L.D. 1999. An Agnps-based runoff and sediment yield model for two small watersheds in Germany. Transactions of the ASAE. VOL. 42(6): 1723-1731

Gupta, H.V., S. Sorooshian, P.O. Yapo. 1998. Toward Improved Calibration of Hydrologic Models: Multiple and Noncommensurable Measures of Information. Water Resources Research 34(4): 751-763.

Haith, D.A., Shoemaker, L.L. 1987, Generalized watershed loading functions for stream flow nutrients: Water Resources Bulletin, v. 23, no. 3, p. 471-478.

Hargreaves, G.H. 1974. "Estimation of Potential and Crop Evapotranspiration," Transactions ASAE, Vol. 17: 701-704.

Heatwole, C.D., Campbell, K.L., Bottcher, A.B. 1987. Modified CREAMS Hydrology Model for Coastal Plain Flatwoods, Transactions of the ASAE 30(4):1014-1022.

Heatwole, K.K., McCray, J.E. 2007. Modeling potential vadose-zone transport of nitrogen from onsite wastewater systems at the development scale, J. Contaminant Hydrology, 91, (1-2), 184-201.

He, C., Riggs, J., Kang, Y. 1993: Integration of geographic information systems and a computer model to evaluate impacts of agricultural runoff on water quality. Water Resources Bulletin, 29 (6), p 891-900.

Herr J., Weintraub, L., Chen, C.W.. 2001. User's Guide to WARMF. Documentation of Graphical User Interface. Prepared for Electric Power Research Institute, Palo Alto, California.

Hill, M.C., Tiedeman, C.R. 2007. Effective groundwater model calibration, with analysis of sensitivities, predictions, and uncertainty: New York, Wiley and Sons, 455 p.

Hill, M.C. 1998, Methods and guidelines for effective model calibration: U.S. Geological Survey Water-Resources Investigations Report 98-4005, 90 p.

Hill, M.C. 1992: A Computer Program (MODFLOWP) for Estimating Parameters of a Transient, Three-dimensional, Ground-water Flow Model Using Nonlinear Regression. Open File Report 91-484, USGS.

Hinkle, S.R., Böhlke, J.K., Duff, J.H., Morgan, D.S., Weick, R.J. 2007, Aquifer scale controls on the distribution of nitrate and ammonium in groundwater near La Pine, Oregon, USA: Journal of Hydrology, v. 333, p. 486-503.

Hinkle, S.R., Morgan, D.S., Orzol, L.L., Polette, D.J. 2007, Groundwater redox zonation near La Pine, Oregon—Relation to river position within the aquifer–riparian zone continuum: U.S. Geological Survey Scientific Investigations Report 2007-5239, 28 p.; last accessed November 9, 2007, at http://pubs.usgs.gov/sir/2007/5239/.

Hinkle, S.T., Weick, R.J., Johnson, J.M., Cahill, J.D., Smith , S.G., Rich. B.J. (2005). Organic wastewater compounds, pharmaceuticals, and coliphage in groundwater receiving discharge from onsite wastewater treatment systems near La Pine, Oregon: Occurrence and implications for transport. *U.S. Geological Survey Scientific Investigations Report 2005* 5055: 98.

Hofstra, W., Hall, D. 1975. Geologic control of supply and quality of water in the mountainous part of Jefferson County, Colorado. Colorado Geological Survey Bulletin 36, 51p.

Holland, J.H. 1975. Adaptation in Natural and Artificial Systems. University of Michigan Press, Ann Arbor, MI. 183 pp.

Holmberg, H., Dilks, D., John, S,, White, D.1998: Modeling water quality response to fertilizer BMPs. Water resources engr '98 (S. Abt, J. Young-Pezeshk & C. Watson,eds.), 2, pp. 1171-1176. New York: ASCE.

Howarth, R.W., Marino, R. 2006. Nitrogen as the limiting nutrient for eutrophication in coastal marine ecosystems: Evolving views over three decades. Limnology and Oceanography, 51(1): 364-376.

Iman, R.L., Conover, W.J. 1980. Small sample sensitivity analysis techniques for computer models, with an application to risk assessment. Communications in Statistics: Theory and Methods A9 (17), 1749-1842

Isaaks, E.H., Srivastava, R.M. 1989. An Introduction to Applied Geostatistics, Oxford University Press, New York, NY.

Jia, Y.B., Culver, T.B. (2006), Robust optimization for total maximum daily load allocations, Water Resources Research, 42(2), W02412.

Johansen, N.B., Imhoff, J.C., Kittle, J.C., Donigian, A.S., 1984,Hydrological Simulation Program - FORTRAN (HSPF): Users Manual Release 8, EPA-600/3-84-066, USEPA, Athens, GA.

KCM, Inc. 1997. South County Regional Cost/Benefit Analysis: Regional Problem Solving Project Final Report. KCM, Inc., Portland, OR.

Kellogg D.Q, Joubert, L., Gold, A, 2001. MANAGE: A Method for Assessment, Nutrient-loading, and Geographic Evaluation of Non-Point Pollution. University of Rhode Island, Department of Natural Resources Science.

Kellogg, D.Q., Joubert, L., Gold, A. 1996, Technical documentation, nutrient loading component of the MANAGE Geographic Information System-based risk assessment method: University of Rhode Island Cooperative Extension.

Khu, S.T., Madsen H, 2005. Multiobjective Calibration with Pareto Preference Ordering: An Application to Rainfall-Runoff Model Calibration. Water Resources Research 41(3): W03004.

Kleinman, P.J., Sharpley, A.N., Wolf, A.M., Beegle, D.B., Moore, P.A. 2002. Measuring water-extractable phosphorus in manure as an indicator of phosphorus in runoff. Soil Science Society of America Journal, 66: 2009-2015.

Knisel, W., Williams, J. 1995: Hydrology components of CREAMS and GLEAMS models. In: Computer models of watershed hydrology (V.P. Singh, ed.), Chap. 28, pp. 1069-1114. Highlands Ranch, CO: Water Resources Publications.

Knisel, W.G., ed. 1980. CREAMS: A Field-Scale Model for Chemical, Runoff, and Erosionfrom Agricultural Management Systems. Conservation Research Report 26, U.S. Department of Agriculture, Washington, D.C.

Kuczera, G., Parent, E. (1998), Monte Carlo assessment of parameter uncertainty in 830 conceptual catchment models: the Metropolis algorithm, *Jour. of Hydrol.,* 211:831 69-85.

Legates, D.R., McCabe G.J. 1999. Evaluating the use of "goodness-of-fit" measures in hydrologic and hydroclimatic model validation. Water Resources Res. 35(1): 233-241.

Lemonds, P.J., McCray, J.E. 2003. Modeling Phosphorus Transport in the Blue River Watershed, Summit County, Colorado. The First Interagency Conference on Research in Watersheds, 27-30 October 2003, Benson, AZ.

Leonard, R.A., Knisel, W.G., Still, D.A., 1987. GLEAMS: Groundwater Loading Effects of Agricultural Management Systems. Trans. ASAE 30:1403-1418.

Leonard, R., Truman, C., Knisel, W., Davis, F. 1992: Pesticide runoff simulations: Long-term annual means vs. event extremes? *Weed Technology, 6* (3), pp. 725-730.

León, L.F., Lam, D.C. (2000), The AGNPS Watershed Model, Chapter 6 in Preparation of a Preliminary Physical Processes and Water Quality Model for Lake Malawi/Nyasa, Final Report, UNU/INWEH, WL Delft Hydraulics, UW, March 2000, pp. 6.1-6.20.

León, L.F., Bowen, G.S., Booty, W.G. 2004. Verification of the AGNPS model trace source contribution algorithm for use in southern Ontario watersheds. IEMSs University of Osnabrück, Germany.

Lin Z., Radcliffe, D.E. 2006. Automatic Calibration and Predictive Uncertainty Analysis of a Semi distributed Watershed Model. Vadose Zone Journal 5:248-260.

Litke, D.W. 1999. Review of phosphorus control measures in the United States and their effects on water quality. In: W.R.I.R.-. U.S. Geological Survey (Editor). National Water Quality Assessment Program.

Lodwig E.M., Hosie, A.H.F., Bordes, A., Findlay, K., Allaway, D., Karunakaran, R., Downie, J.A., Poole, PS., 2003. Amino-acid cycling drives nitrogen fixation in the legume—*Rhizobium* symbiosis. Nature 422: 722-726.

Mandel, R., Haith. D.A., 1992. The Impact of Septic Systems on Surface Water Quality. Dept. of Civil and Environmental Engineering, Dept. of Agriculture and Biological Engineering, Cornell University.

Martin, J.P., Focht, D.D. 1977. Soils for managinment of organic wastes and wastewaters. In: L.F. Elliott and F.J. Stevenson (Editors). Soil Science Society of America, American Society of Agronomy, Crop Science Society of America, Madison, WI, pp. 650.

Maurer, M., Abramovich, D., Siegrist, R.L., Gujer, W. 1999. Kinetics of biologically induced phosphorus precipitation in waste-water treatment. Water Research, 33(2): 484-493.

McCray, J.E., Kirkland, S.L., Siegrist, R.L., Thyne, G.D. 2005. Model parameters for simulating fate and transport of onsite wastewater nutrients. Ground Water, 43(4): 628-639.

McIntosh, T.H.; Qiu, Huaying; Harris, H. 1993. Use of agricultural computer simulation models for water quality-land management decisions in N.E. Wisconsin. Ome 36th Conference of the International Association for Great Lakes Research, June 4-10, 1993. Program and Abstracts.; P. 118.

McKay, M.D., Beckman, R.J., Conover, W.J. 1979. A comparison of three methods for selecting values of input variables in the analysis of output from a computer code. Technometrics 21 (2), 239-245.

McKay, M.D. 1988. In: Ronen, Y. (Ed.), Sensitivity and Uncertainty Analysis Using a Statistical Sample of Input Values. CRC Press, Boca Raton, FL, pp. 145-186.

McElroy, A.D., Chiu, S.Y., Nebgen, J.W. 1976. Load functions for assessment of water pollution from non-point sources. EPA document EPA 600/2-76-151. USEPA, Athens, GA

Mengis, M., Harris, M., English, M.C., Aravena, R., Elgood, R.J., MacLean, A. 1999. Multiple geochemical and isotopic approaches for assessing groundwater NO_3^- elimination in a riparian zone. Ground Water, 37(3): 448-457.

Menzel, R.G. 1980. Enrichment ratios for water quality modeling. p. 486-492. In W.G. Knisel (ed.) CREAMS, A field scale model for chemicals, runoff, and erosion from agricultural management systems. U.S. Dept. Agric. Conserv. Res. Rept. No. 26.

Miettinen, I.T., Vartiainen, T., Martikainen, P.J. 1997. Phosphorus and bacterial growth in drinking water. Applied and Environmental Microbiology, 63(8): 3242-3245.

Mitchell, J., Engel, B., Srinivasan, R., Wang, S., 989: Validation of AGNPS for small watersheds using an integrated AGNPS/GIS system. Conf. on Ionic Currents in Development, pp. 833-842. Biological Bulletin, Marine Biological Laboratory, Woods Hole, 176(2) Suppl.

Motovilov, Y.G., Gottschalk, L., Engeland, K., Rodhe, A. 1999. Validation of a distributed hydrological model against spatial observations. Agric. and Forest Meteorology 98-99:257-277.

Morgan, D. S., Everett, R. 2005. Simulation-Optimization Methods for Management of Nitrate Loading to Groundwater from Decentralized Wastewater Treatment Systems. Project No. WU-HT-03-37. Prepared for the National Decentralized Water Resources Capacity Development Project, Washington University, St. Louis, MO, by U.S. Geological Survey, Oregon Water Science Center, Portland, OR.

Morgan, K. 2000. Spatial Analysis and Modeling of Geochemical Distribution to AssessFracture Flow in Turkey Creek Basin, Jefferson County, Colorado. Colorado School of Mines MS thesis. 194 p.

Morgan, D.S., Hinkle, S.R., Weick, R.J. 2007, Evaluation of approaches for managing nitrate loading from onsite wastewater systems near La Pine, Oregon: U.S. Geological Survey Scientific Investigations Report 2007-5237, 64 p.; Available at http://pubs.usgs.gov/sir/2007/5237/

Moriasi, D.N., Arnold, J.G., Van Liew, M.W., Bingner, R.L., Harmel, R.D., Veith, T.L. 2007. Model evaluation guidelines for systematic quantification of accuracy in watershed simulations. Transactions of the ASABE. 50(3):885-900.

Muleta, M.K., Nicklow J.W. 2005. Sensitivity and uncertainty anaylsis coupled with automatic calibration for a distributed watershed model. J. of Hydrology, 306 (2005):127-145.

Muller, S., Gregory, J. 2003. Groundwater Loading Effects of Agricultural Management Systems Version 3.0 Calibration and Verification of the GLEAMS model for Buck Island Ranch University of Florida Department of Agricultural and Biological Engineering.

Murray, K.E., Straud, D.R., Hammond, W.W. 2007. Characterizing groundwater flow in a faulted karst system using optical brighteners from septic systems as tracers: Environmental Geology, v. 53, p. 769-776.

Nash, J.E., Sutcliffe, J.V. 1970. River flow forecasting through conceptual models, part 1 - A discussion of principles, J. of Hydrology 10(3):282-290.

NCDC, 2007. The National Oceanic and Atmospheric Administration (NOAA). National Climatic Data Center. US Department of Commerce. http://www.ncdc.noaa.gov/oa/climate/stationlocator.html

Neary, D., Bush, P., Michael, J. 1993: Fate, Dissipation and Environmental Effects of Pesticides in Southern Forests: A Review of a Decade of Research Progress. Environmental Toxicology and Chemistry, 12 (3), pp. 411-428.

Neitsch, S.L., Arnold, J.G. Kiniry, J.R. Williams, J.R., King, K.W. 2002. Soil and Water Assessment Tool (SWAT). Theoretical documentation, version 2000. Water Resources Institute, College Station, Texas, TWRI Report TR-192.

Neitsch, S.L., Arnold, J.G., Kiniry, J.R., Srinivasan, R.,Williams J.R. 2004. Soil and Water Assessment Tool (SWAT). Input/Output File Documentation , version 2005.

Neitsch, S.L., Arnold, J.G., Kiniry, J.R., Williams, J.R. 2005. Soil and Water Assessment Tool (SWAT). Theoretical documentation, version 2005. http://www.brc.tamus.edu/swat/downloads/doc/swat2005/SWAT%202005%20theory%20final.pdf.

Newton, B.J., Jarrell, W.M. 1999. A procedure to estimate the response of aquatic systems to changes in phosphorus and nitrogen inputs. In: U.D.o.A. National Water and Climate Center, Natural Resources Conservation Service (Editor).

Ogden, F.L., Julien, P.Y. 2002: CASC2D: A Two-Dimensional, Physically-Based, Hortonian, Hydrologic Model, in Mathematical Models of Small Watershed Hydrology and Applications, (V.J. Singh, and D. Freverts, eds.), Water Resources Publications, Littleton, Colorado, 972 pp.

Orzol, L., McGrath, T. 1989: Summary of modifications of the U.S. Geological Surveymodular, finite-difference, ground-water flow model to read and write geographic information system files. Biological Bulletin, 176 (2) Suppl. Marine Biological Laboratory, Woods Hole.

Otis, R.N., Hutelek, N.J., Boyle, W.C. 1973, "Onsite Household Wastewater Treatment Alternatives, Laboratory and Field Studies," paper, presented at the Rural Environmental Engineering Conference, Warren, VT.

Panuska, J.C., Moore, I.D., Kramer, L.A. 1991: Terrain analysis: Integration into the agricultural non-point source (AGNPS) pollution model. Journal of Soil and Water Conservation, 46(1), pp. 59-66.

Parry, S., Renault, P., Chadeuf, J., Chenu, C., Lensi, R. 2000. Particulate organic matter as a source of variation in denitrification in clods of soil. European Journal of Soil Science, 51(2): 271-281.

Perkins, S.P; Sophocleous, M, 1999. Development of a Comprehensive Watershed Model Applied to Study Stream Yield under Drought Conditions, Ground Water, 37(3), 418-426.

Poeter, E.P., Hill, M.C. 1997, Inverse Methods: A Necessary Next Step in Groundwater Modeling, Ground Water, v. 35, no. 2, pp. 250-260.

Poeter E.P, Hill M.C., Banta E.R., Mehl S., Christensen S. 2005. UCODE_2005 and Six Other Computer Codes for Universal Sensitivity Analysis, Calibration, and Uncertainty Evaluation: U.S. Geological Survey Techniques and Methods 6-A11(2005) 283 pp.

Poeter, E.P., Anderson, D. 2005, Multi-model Ranking and Inference in Ground-Water Modeling, Ground Water v. 43, no. 4: 597-605.

Powell, C.L.L., Daniel, J. 1978. Mycorrhizal fungi stimulate uptake of soluble and insoluble phosphate fertilizer from a phosphate-deficient soil. New Phytologist, 80(2): 351-358.

Pradhan, S., Hoover, T., Devine, H.A., Arnold, J.G., Di Luzio. M.2005. Quantifying Nitrogen Exports from Onsite Systems to Surface Waters within Small Watersheds using SWAT Model. Proceeding of the Third Conference on Watershed Management to Meet Water Quality Standards and Emerging TMDL. Atlanta, GA. March 2005, ASAE :387-396

Refsgaard, J.C., Knudsen, J. 1996. Operational validation and intercomparison of different types of hydrologic models. Water Resources and Research 32 (7), 2189-2202.

Refsgaard, J.C. 1997. Parameterization, calibration and validation of distributed hydrologic models. Journal of Hydrology 198, 69-97.

Rezania, L. 2004. Optimizing phosphate treatment to minimize lead/copper seasonal variations, Waterline. Minnesota Department of Health, Environmental Health Division.

Robertson, W.D., Schiff, S.L., Ptacek, C.J. 1998. Review of phosphate mobility and persistence in 10 septic system plumes. Ground Water, 33(6): 1000-1010.

Robertson, W.D., Cherry, J.A. 1995. In situ denitrification of septic system nitrate using reactive porous media barriers. Field Trials. Ground Water, 33(1): 99-111.)

Ryther, J.H., Dunstan, W.M. 1971. Nitrogen, phosphorus and eutrophication in the coastal marine environment. Science, 171: 1008-1013.

Saltelli, Andres, A., Homa, T.H. 1995. Sensitivity analysis of model output; performance of the iterated fractional design method. Computational Statistics and Data Analysis 20, 387-407.

Saltelli, A., Chan, K., Scott, E.M. (Eds.), 2000. Sensitivity Analysis. Wiley, New York.

Santhi, C, Arnold, J.G., Williams, J.R., Dugas, W.A., Srinivasan, R., Hauck, L.M. 2001. Validation of the SWAT model on a large river basin with point and nonpoint sources. J. American Water Resources Assoc. 37(5): 1169-1188.

Schindler, D.W. 1977. Evolution of phosphorus limitation in lakes. Science, 195: 260-262.

SCS, 1972, Section 4: Hydrology, National Engineering Handbook: Washington, D.C., USDA Soil Conservation Service.

Sekiguchi, Y., Matsunaga, A., Yamamoto, A., Inoue, Y. 2000. Analysis of condensed phosphates in food products by ion chromatography with an on-line hydroxide effluent generator. Journal of Chromatography A, 881: 639-644.

Senarath, S.U.S., Ogden, F.L., Downer, C.W., Sharif, H.O. 2000. On the calibration and verification of two-dimensional, distributed, hortonian, continuous watershed models. Water Resources and Research 36 (6), 1495-1510.

Sham, C.H., Brawley J.W., Moritz, M.A. 1995. Quantifying Septic Nitrogen Loadings to Receiving Waters: Waquoit Bay, Massachusetts. International Journal of Geographical Information Systems. 9(4): 463-473.

Shirmohammadi, A., Knisel, W. 1994: Evaluation of the GLEAMS model for pesticide leaching in Sweden. J. Env. Sci. Health, Part A, Env. Sci. Engr., A29 (6), pp. 1167-1182.

Shirmohammadi, A., Chaubey, I., Harmel, R.D., Bosch, D.D., Munoz-Carpena, R., Dharmasri, C., Sexton, A., Arabi, M., Wolfe, M.L., Frankenberger, J., Graff, C., Sohrabi, T.M. 2006, Uncertainty in TMDL models: Transactions of the ASABE, v. 49, no. 4, p. 1033-1049.

Shirmohammadi, A, Ulen, B., Bergstrom, L., Knisel, W. 1998: Simulation of nitrogen and phosphorus leaching in a structured soil using gleams and a new submodel, "PARTLE." Trans. ASAE, 41 (2), pp. 353-360

Shuval, H.I., Gruener, N. 1972. Epidemiological and toxicological aspects of nitrates and nitrites in the environment. American Journal of Public Health, 62: 1045-1051.

Simunek, J., Sejna, M., Van Genuchten, M.T. 1999, The HYDRUS-2D software package for simulating two-dimensional movement of water, heat, and multiple solutes in variably saturated media. Version 2.0, International Ground Water Modeling Center: Golden, CO, p. 251.

Singh, J., Knapp, H.V., Demissie M. 2004. Hydrologic modeling of the Iroquois River watershed using HSPF and SWAT. ISWS CR 2004-08. Champaign, Ill.: Illinois State Water Survey. Available at: www.sws.uiuc.edu/pubdoc/CR/ ISWSCR2004-08.pdf.

Skaggs R.W. 1977. Chemicals, Runoff and Erosion from Agricultural Management Systems.General Model Information.

Sloan, P.G., Morre, I.D., Coltharp, G.B., Eigel, J.D. 1983. Modeling Surface and Subsurface Stormflow on Steeply-Sloping Forested Watersheds. Water Resources Institute Report 142, University of Kentucky, Lexington, Kentucky.

Sophocleous, M.A., Koelliker, J.K., Govindaraju, T. 1999. Integrated numerical modeling for basin-wide water management: The case of the Rattlesnake Creek basin in south-central Kansas, J. Hydrol., 214, 179-196.

Spear, R.C., Hornberger, G.M. 1980. Eutrophication in Peel Inlet, II, Identification of critical uncertainties via generalized sensitivity analysis. Water Research 14, 43-49.

Sposito, G. 1989. The Chemistry of Soils. Oxford University Press, New York.

Srinivasan, R, Arnold, J. 1994: Integration of a basin-scale water quality model with GIS. Water Resources Bulletin, 30 (3), pp.453-462.

Stephens, M.A. 1970. Use of the Kolmogorov Smirnov, Cramervon Mises and related statistics without extensive tables. Journal of the Royel Statistical Society: Series B33, 115-122.

Stolt, M., Reneau, R. 1991. Potential for Contamination of Ground and Surface Waters from Onsite Surface Disposal Systems. Virginia Department of Health.

Sugiharto, T, McIntosh, T., Uhrig, R., Lardinois, J. 1994: Modeling alternatives to reduce dairy farm and watershed non-point source pollution. Journal of Environmental Quality, 23 (1), pp. 18-24.

Swain, E., Wexler, E. 1992: A Coupled Surface-Water and Ground-Water Flow Model for Simulation of Stream-Aquifer Interaction. Open-File Report 92-138, USGS.

Tang, C., Kondoh, A. 1996: Regional groundwater model with GIS. 30th International Geological Conference. 1, p. 471.

Tim, U.S. 1996. Coupling vadose zone models to GIS: Emerging trends and bottlenecks. J. Environ. Qual. 25:535-544.

Tucholke, M.B., McCray, J.E., Thyne, G.D., Waskom, R.M. (2007). Correlating Denitrification Rates to Soil Texture using Hierarchical Cluster Analysis. Eleventh National Symposium on Individual and Small Community Sewage Systems, ASABE, October 2007.

Tucholke, M.B. 2007. Statistical Assessment of Relationships Between Denitrificaiton and Easily Measurable Soil Properties: A Simple Predictive Tool for Watershed-Scale Modeling. MS Thesis Hydrologic Science and Engineering, Colorado School of Mines, Golden, CO.

USDA, 2007. Natural Resources Conservation Service (NRCS), National Cartography and Geospatial Center. Soil Survey Geographic (SSURGO). http://www.ncgc.nrcs.usda.gov/products/datasets/statsgo/index.html.

USDA, 1992. STATSGO - State soils geographic database. Soil Conservation Service, Publ. Number 1492, Washington D.C. Ward G.H., J. Benaman, 1999: Models for TMDL application in Texas watercourses: screening and model review. Submitted to Texas Natural Resource Conservation Commission, Austin, TX.

U.S. EPA, 2003, Voluntary national guidelines for management of onsite and clustered (decentralized) wastewater treatment systems, U.S. Environmental Protection Agency, 832-B-03-001.

Valiela, I., Bowen, J.L., Kroeger, K.D. 2002, Assessment of models for estimation of land derived nitrogen loads to shallow estuaries: Applied Geochemistry, v. 17, no. 7, p. 935-953.

Valiela, I., Collins, G., Kremer, J., Lajtha, K., Geist, M., Seely, B., Brawley, J., Sham, C.H. 1997, Nitrogen loading from coastal watersheds to receiving estuaries: New method and application: Ecological Applications, v. 7, no. 2, p. 358-380.

Van der Perk, M., Bierkens, M.F.P. 1997. The identifiability of parameters in a water quality model of the Biebrza River, Poland. Journal of Hydrology 200, 307-322.

Van Griensven A., Meixner, T., Grunwald S., Bishop T., Diluzio M., Srinivasan, R., A global sensitivity analysis tool for the parameters of multi-variable catchment models Journal of Hydrology 324 (2006) 10-23.

Van Liew, M. W., Arnold, J.G., Garbrecht, J.D. 2003. Hydrologic simulation on agricultural watersheds: Choosing between two models. *Trans. ASAE* 46(6): 1539-1551.

Voss, C.I. 1984, A finite-element simulation model for saturated-unsaturated, fluid-density dependent ground-water flow with energy transport or chemically-reactive single-species solute transport, U.S. Geological Survey Water-Resources Investigation Report 84-4369, p. 409.

Vrugt, J.A., Gupta H.V., Bastidas L.A., Bouten W., Sorooshian S. 2003. Effective and Efficient Algorithm for Multiobjective Optimization of Hydrologic Models. Water Resources Research 39(8): 1214.

Weintraub L.H Z., Chen, C.W., Goldstein, R.A., Siegrist, R.L. 2004. WARMF: A Watershed Modeling Tool For Onsite Wastewater Systems. In Proceedings of The Tenth National Symposium on Individual and Small Community Sewage Systems, 636-646. ASAE. St. Joseph, MI. Presented March 21-24, 2004, Sacramento, CA.

Weiskel, P.K., Howes, B.L. 1992. Differential transport of sewage-derived nitrogen and phosphorus through a coastal watershed. Environmental Science and Technology, 26(2): 352-360.)

Whelan, B.R. 1988. Disposal of septic tank efluent in calcareous sands. J. Environmental Quality, 17(2): 272-277.

Wehrmann, H.A. 1984. "Managing Ground Water Nitrate Quality by Mass Balance Modeling in the Rockton-Roscoe Area, Illinois." Proceedings of the NWWA Eastern Regional Conference on Ground Water Management. National Water Well Association, Dublin, OH.

Wilhelm, S.R., Schiff, S.L., Robertson, W.D. 1994. Chemical Fate and Transport in a Domestic Septic System: Unsaturated and Saturated Zone Geochemistry. Environmental Toxicology and Chemistry, 13, 2: 193-203.

Williams, J.R., Hann, R.W. 1972. HYMO, a problem-oriented computer language for building hydrologic models. Water Resour. Res. 8(1):79-85.

Willmott, C.J. 1981. On the validation of models. Physical Geography 2: 184-194.

Wischmeier, W.H., Smith D.D., 1978. Predicting Rainfall Erosion Losses. U.S. Department of Agriculture, Agriculture Handbook No. 537, 58pp.

Wood, J.S.A. 1993. Migration of septic system contaminated groundwater to a lake in a Precambrian shield setting: A case study. M.Sc. thesis, University of Waterloo, Waterloo, Ontario.

Yacob, S. 2004. Using Multivariate Statistical Analyses to Characterize the Effects of Population Growth on Water Quality in a Mountain Watershed. Colorado School of Mines MS thesis. 108 p.

Yan, J., Smith, K. 1994: Simulation of integrated surface water and groundwater systems -Model formulation. Water Resources Bulletin, 30 (5), pp. 879-890.

Yalin, Y.S. 1963. An expression for bed-load transportation. J. Hydraulics Div., proc. Am. Soc. Civil Eng. 89(HY3): 221-250.

Yapo, P.O., Gupta, H.V., Sorooshian, S. 1998. Multi-objective Optimization for Hydrologic Models. Journal of Hydrology 204(1): 83-97.

Yen, B.C., Cheng, S.-T., Melching, C.S. 1986. First order reliability analysis: stochastic and risk analysis in hydraulic engineering. In: Yen, B.C. (Ed.), Water Resources Publications, Littleton, CO, pp. 1-36.

Yoon, K., Yoo, K., Wood, C., Hall, B. 1994: Application of GLEAMS to predict nutrient losses from land application of poultry litter. *Trans. Am. Soc. Ag. Engr., 37* (2), pp. 453-459.

Young, R.A., Onstad, C.A, Bosch, D.D., Anderson, W.P. 1995. AGNPS: An agricultural non-point source model, in Singh V.P. (ed.), Computer models of watershed hydrology: Highlands ranch, Colorado, Water Resources Publications. 1011-1020.

Yuan, Y., Binger. R.L., Rebich, R.A. 2001. Evaluation of AnnAGNPS on Mississippi Delta MSEA Watersheds. Trans. ASAE. 44(5): 1183-1190.

Yuan, Y, Binger. R.L., Rebich, R.A. 2003. Evaluation of AnnAGNPS Nitrogen Loading in an Agricultural Watershed. J. Am. Wat. Resources Assn. 39(2): 457-466.

Zacharias, S., Heatwole, C.D., Persaud, N., Bruggeman, A.C., Kumar, D., Smith. C.N. 1999. Stochastic simulation of fi eld-scale pesticide transport using Opus and GLEAMS. J. Environ. Qual. 28:411-423.

Zanini, L., Robertson, W.D. Ptacek, C.J. Schiff , S.L. Mayer T. 1998. "Phosphorus characterization in sediments impacted by septic effluent at four sites in central Canada." Journal of Contaminant Hydrology 33: 405-429.

Zhang, D. 2002. Stochastic Methods for Flow in Porous Media: Coping with Uncertainties, Academic Press, San Diego, CA.

Zheng, C. 1993. A modular three-dimensional transport model for simulation of advection, dispersion and chemical reaction of contaminants in groundwater systems. United States Environmental Protection Agency Report.

Zheng, C., Wang, P.P. 1999, MT3DMS, A modular three-dimensional multi-species transport model for simulation of advection, dispersion and chemical reactions of contaminants in groundwater systems; documentation and user's guide, U.S. Army Engineer Research and Development Center Contract Report SERDP-99-1, Vicksburg, MS, 202 p.

Zimmerman, M.J. 2004. Occurrence of Organic Wastewater Contaminants, Pharmaceuticals, and Personal Care Prodcuts in Selected Water Supplies, Cape Cod, Massachusetts. *U.S. Geological Survey Scientific Open-File Report 2005* 1206: 16.

Zurawski, M.A., Robertson, W.D., Ptacek, C.J., Schiff, S.L. 2004. Geochemical stability of phosphorus solids below septic system infiltration beds. Journal of Contaminant Hydrology, 73: 129-143.

WERF SUBSCRIBERS

WASTEWATER UTILITY

Alabama
Montgomery Water Works & Sanitary Sewer Board

Alaska
Anchorage Water & Wastewater Utility

Arizona
Avondale, City of
Glendale, City of, Utilities Department
Mesa, City of
Peoria, City of
Phoenix Water Services Dept.
Pima County Wastewater Management
Safford, City of
Tempe, City of

Arkansas
Little Rock Wastewater Utility

California
Central Contra Costa Sanitary District
Corona, City of
Crestline Sanitation District
Delta Diablo Sanitation District
Dublin San Ramon Services District
East Bay Dischargers Authority
East Bay Municipal Utility District
El Dorado Irrigation District
Fairfield-Suisun Sewer District
Fresno Department of Public Utilities
Inland Empire Utilities Agency
Irvine Ranch Water District
Las Gallinas Valley Sanitary District
Las Virgenes Municipal Water District
Livermore, City of
Los Angeles, City of
Los Angeles County, Sanitation Districts of
Napa Sanitation District
Novato Sanitary District
Orange County Sanitation District
Palo Alto, City of
Riverside, City of
Sacramento Regional County Sanitation District
San Diego Metropolitan Wastewater Department, City of
San Francisco, City & County of
San Jose, City of
Santa Barbara, City of
Santa Cruz, City of
Santa Rosa, City of
South Bayside System Authority
South Coast Water District
South Orange County Wastewater Authority
Stege Sanitary District
Sunnyvale, City of
Union Sanitary District
West Valley Sanitation District

Colorado
Aurora, City of
Boulder, City of
Greeley, City of
Littleton/Englewood Water Pollution Control Plant
Metro Wastewater Reclamation District, Denver

Connecticut
Greater New Haven WPCA
Stamford, City of

District of Columbia
District of Columbia Water & Sewer Authority

Florida
Broward, County of
Fort Lauderdale, City of
Jacksonville Electric Authority (JEA)
Miami-Dade Water & Sewer Authority
Orange County Utilities Department
Pinellas, County of
Reedy Creek Improvement District
Seminole County Environmental Services
St. Petersburg, City of
Tallahassee, City of
Tampa, City of
Toho Water Authority
West Palm Beach, City of

Georgia
Atlanta Department of Watershed Management
Augusta, City of
Clayton County Water Authority
Cobb County Water System
Columbus Water Works
Fulton County
Gwinnett County Department of Public Utilities
Savannah, City of

Hawaii
Honolulu, City & County of

Idaho
Boise, City of

Illinois
Decatur, Sanitary District of
Greater Peoria Sanitary District
Kankakee River Metropolitan Agency
Metropolitan Water Reclamation District of Greater Chicago
Wheaton Sanitary District

Iowa
Ames, City of
Cedar Rapids Wastewater Facility
Des Moines, City of
Iowa City

Kansas
Johnson County Wastewater
Lenexa, City of
Unified Government of Wyandotte County/ Kansas City, City of

Kentucky
Louisville & Jefferson County Metropolitan Sewer District

Louisiana
Sewerage & Water Board of New Orleans

Maine
Bangor, City of
Portland Water District

Maryland
Anne Arundel County Bureau of Utility Operations
Howard County Bureau of Utilities
Washington Suburban Sanitary Commission

Massachusetts
Boston Water & Sewer Commission
Massachusetts Water Resources Authority (MWRA)
Upper Blackstone Water Pollution Abatement District

Michigan
Ann Arbor, City of
Detroit, City of
Holland Board of Public Works
Saginaw, City of
Wayne County Department of Environment
Wyoming, City of

Minnesota
Rochester, City of
Western Lake Superior Sanitary District

Missouri
Independence, City of
Kansas City Missouri Water Services Department
Little Blue Valley Sewer District
Metropolitan St. Louis Sewer District

Nebraska
Lincoln Public Works and Utilities Department

Nevada
Henderson, City of
Las Vegas, City of
Reno, City of

New Jersey
Bergen County Utilities Authority
Ocean County Utilities Authority
Passaic Valley Sewerage Commissioners

New York
New York City Department of Environmental Protection

North Carolina
Charlotte/Mecklenburg Utilities
Durham, City of
Metropolitan Sewerage District of Buncombe County
Orange Water & Sewer Authority
University of North Carolina, Chapel Hill

Ohio
Akron, City of
Butler County Department of Environmental Services
Columbus, City of
Metropolitan Sewer District of Greater Cincinnati
Montgomery, County of
Northeast Ohio Regional Sewer District
Summit, County of

Oklahoma
Oklahoma City Water & Wastewater Utility Department
Tulsa, City of

Oregon
Albany, City of
Clean Water Services
Eugene, City of
Gresham, City of
Portland, City of Bureau of Environmental Services
Water Environment Services

Pennsylvania
Hemlock Municipal Sewer Cooperative (HMSC)
Philadelphia, City of
University Area Joint Authority

South Carolina
Charleston Water System
Mount Pleasant Waterworks & Sewer Commission
Spartanburg Water

Tennessee
Cleveland Utilities
Murfreesboro Water & Sewer Department
Nashville Metro Water Services

Texas
Austin, City of
Dallas Water Utilities
Denton, City of
El Paso Water Utilities

WERF SUBSCRIBERS

Fort Worth, City of
Houston, City of
San Antonio Water System
Trinity River Authority

Utah
Salt Lake City Corporation

Virginia
Alexandria Sanitation Authority
Arlington, County of
Fairfax County
Hampton Roads Sanitation District
Hanover, County of
Henrico, County of
Hopewell Regional Wastewater Treatment Facility
Loudoun Water
Prince William County Service Authority
Richmond, City of
Rivanna Water & Sewer Authority

Washington
Everett, City of
King County Department of Natural Resources
Seattle Public Utilities
Sunnyside, Port of
Yakima, City of

Wisconsin
Green Bay Metro Sewerage District
Kenosha Water Utility
Madison Metropolitan Sewerage District
Milwaukee Metropolitan Sewerage District
Racine, City of
Sheboygan Regional Wastewater Treatment
Wausau Water Works

Australia
ACTEW (Ecowise)
South Australian Water Corporation
South East Water Limited
Sydney Water Corporation
Water Corporation of Western Australia

Canada
Edmonton, City of/Edmonton Waste Management Centre of Excellence
Lethbridge, City of
Regina, City of, Saskatchewan
Toronto, City of, Ontario
Winnipeg, City of, Manitoba

New Zealand
Watercare Services Limited

United Kingdom
Anglian Water Services, Ltd.

STORMWATER UTILITY

California
Fresno Metropolitan Flood Control District
Los Angeles, City of, Department of Public Works
Monterey, City of
San Francisco, City & County of
Santa Rosa, City of
Sunnyvale, City of

Colorado
Aurora, City of
Boulder, City of

Florida
Orlando, City of

Iowa
Cedar Rapids Wastewater Facility
Des Moines, City of

Kansas
Overland Park, City of

Kentucky
Louisville & Jefferson County Metropolitan Sewer District

Maine
Portland Water District

North Carolina
Charlotte, City of, Stormwater Services

Pennsylvania
Philadelphia, City of

Tennessee
Chattanooga Stormwater Management

Texas
Harris County Flood Control District, Texas

Washington
Bellevue Utilities Department
Seattle Public Utilities

STATE

Connecticut Department of Environmental Protection
Kansas Department of Health & Environment
New England Interstate Water Pollution Control Commission (NEIWPCC)
Ohio River Valley Sanitation Commission
Urban Drainage & Flood Control District, CO

CORPORATE

ADS Environmental Services
Advanced Data Mining International
AECOM
Alan Plummer & Associates
Alpine Technology Inc.
Aqua-Aerobic Systems Inc.
Aquateam–Norwegian Water Technology Centre A/S
ARCADIS
Associated Engineering
Bernardin Lochmueller & Associates
Black & Veatch
Blue Water Technologies, Inc.
Brown & Caldwell
Burgess & Niple, Ltd.
Burns & McDonnell
CABE Associates Inc.
The Cadmus Group
Camp Dresser & McKee Inc.
Carollo Engineers Inc.
Carpenter Environmental Associates Inc.
CET Engineering Services
CH2M HILL
CRA Infrastructure & Engineering
CONTECH Stormwater Solutions
D&B/Guarino Engineers, LLC
Damon S. Williams Associates, LLC
Ecovation
EMA Inc.
Environmental Operating Solutions, Inc.
Environ/The ADVENT Group, Inc.
Fay, Spofford, & Thorndike Inc.
Freese & Nichols, Inc.
ftn Associates Inc.
Gannett Fleming Inc.
Garden & Associates, Ltd.
Geosyntec Consultants
GHD
Greeley and Hansen LLC
Hazen & Sawyer, P.C.
HDR Engineering Inc.
HNTB Corporation
Hydromantis Inc.
HydroQual Inc.
Infilco Degremont Inc.
Jacques Whitford NAWE, Inc.
Jason Consultants LLC Inc.
Jordan, Jones, & Goulding Inc.
KCI Technologies Inc.
Kelly & Weaver, P.C.
Kennedy/Jenks Consultants
KMK Consultants
Larry Walker Associates
Limno-Tech Inc.
The Low Impact Development Center Inc.
Malcolm Pirnie Inc.
Material Matters
McKim & Creed
MPR Engineering Corporation, Inc.
MWH
NTL Alaska, Inc.
O'Brien & Gere Engineers Inc.
Odor & Corrosion Technology Consultants Inc.
Paradigm Environmental Technologies, Inc.
Parametrix Inc.
Parsons
Post, Buckley, Schuh & Jernigan
Praxair, Inc.
Ring Industrial Group
RMC Water & Environment
Ross & Associates Ltd.
SAIC
Siemens Water Technologies
The Soap & Detergent Association
Smith & Loveless, Inc.
Southeast Environmental Engineering, LLC
Stearns & Wheler, LLC
Stone Environmental Inc.
Stratus Consulting Inc.
Synagro Technologies Inc.
Tetra Tech Inc.
Trojan Technologies Inc.
Trussell Technologies, Inc.
Uni-Bell PVC Pipe Association
URS Corporation
Westin Engineering Inc.
Wright Water Engineers
Zoeller Pump Company

INDUSTRY

American Electric Power
American Water
Anglian Water Services, Ltd.
Chevron Energy Technology
The Coca-Cola Company
Dow Chemical Company
DuPont Company
Eastman Chemical Company
Eli Lilly & Company
Johnson & Johnson
Merck & Company Inc.
Procter & Gamble Company
Suez Environnment
United Utilities North West (UUNW)
United Water Services LLC

Note: List as of 2/21/09

IWA Publishing's authorised EU representative for General Product Safety Regulations is Diane D'Arras, 15 rue Duret, 75116 Paris, France, e-mail: safety@iwap.co.uk.

Printed and bound by CPI Group (UK) Ltd, Croydon, CR0 4YY
12/05/2026
02108876-0006